Ergebnisse der Mathematik und ihrer Grenzgebiete

Band 45

Herausgegeben von
P. R. Halmos · P. J. Hilton · R. Remmert · B. Szőkefalvi-Nagy

Unter Mitwirkung von
L. V. Ahlfors · R. Baer · F. L. Bauer · R. Courant · A. Dold
J. L. Doob · S. Eilenberg · M. Kneser · G. H. Müller · M. M. Postnikov ·
H. Rademacher · B. Segre · E. Sperner

Geschäftsführender Herausgeber: P. J. Hilton

Yu. V. Linnik

Ergodic Properties of Algebraic Fields

Translated by M. S. Keane

Springer-Verlag New York Inc. 1968

Professor Dr. Yu. V. Linnik
Matematiceski Institut Akademii Nauk SSSR
Leningrad – U.d.S.S.R.

Dr. Michael S. Keane
Mathematisches Institut der Universität Erlangen

Original title: Ergoditscheskie svoistva algebraitscheskich polei. Published by Izdatelstvo Leningradskogo Universiteta, 1967

ISBN-13: 978-3-642-86633-3 e-ISBN-13: 978-3-642-86631-9
DOI: 10.1007/978-3-642-86631-9

 Library of Congress Catalog Card Number 68-19463.

Softcover reprint of the hardcover 1st edition 1968

Title No. 4589

Preface

The applications of ergodic theory to metric number theory are well known; part of the latter theory turns out to be essentially a special case of general ergodic theorems.

In the present book other applications of ergodic concepts are presented. Constructing "flows" of integral points on certain algebraic manifolds given by systems of integral polynomials, we are able to prove individual ergodic theorems and mixing theorems in certain cases. These theorems permit asymptotic calculations of the distributions of integral points on such manifolds, and we arrive at results inaccessible up to now by the usual methods of analytic number theory. Typical in this respect is the theorem concerning the asymptotic distribution and ergodic behavior of the set of integral points on the sphere

$$x^2+y^2+z^2=m$$

for increasing m. It is not known up until now how to obtain the simple and geometrically obvious regularity of the distribution of integral points on the sphere other than by ergodic methods.

Systems of diophantine equations are studied with our method, and flows of integral points introduced for this purpose turn out to be closely connected with the behavior of ideal classes of the corresponding algebraic fields, and this behavior shows certain ergodic regularity in sequences of algebraic fields. However, in this book we examine in this respect only quadratic fields in sufficient detail, studying fields of higher degrees only in chapter VII.

The theory presented here arose from the works of LINNIK [1] in 1940 on positive ternary quadratic forms (at that time the basic ergodic character of the phenomena dealt with was not yet realized). Many significant results in this region have been obtained by A. V. MALISHEV and B. F. SKUBENKO [2].

The reader is assumed to possess elementary knowledge of measure theory, probability theory, algebraic number theory, and the theory of matrices.

Leningrad, June 1967 YU. V. LINNIK

Contents

Introduction

§ 1. Ergodic Theory

Ergodic theory, originating in the classical works of W. GIBBS and A. POINCARÉ, was in the first instance concerned with problems of statistical physics. Not until recent times was the need felt for generalizations of classical dynamical systems. Let us consider a system with n degrees of freedom, described by the generalized coordinates $q_1, \ldots, q_n$ and generalized momenta $p_1, \ldots, p_n$. The total energy of the system is given by the Hamiltonian function $H(q_1, \ldots, q_n, p_1, \ldots, p_n)$ and its motion by the system of differential equations:

$$\frac{dq_i}{dt} = \frac{\partial H}{\partial p_i} \qquad (i=1,2,\ldots,n)$$

$$\frac{dp_i}{dt} = -\frac{\partial H}{\partial q_i} \qquad (i=1,2,\ldots,n).$$

Observe the trajectories of points on the energy surface $H(q_1, \ldots, q_n, p_1, \ldots, p_n) = const.$ Under the condition of compactness of this surface and a few "regularity" conditions, the given equations define a transformation on the energy surface which preserves a measure introduced with the help of the gradient of H. Ergodic theory is concerned with the fraction of time which points, moving according to the transformation described above, spend in a given measurable region of the energy surface during a sufficiently long time interval.

A precise mathematical proof of a statement of this type was first given by G. D. BIRKHOFF in 1931; here the statement of the problem was sufficiently general. A sufficiently detailed resumé of the elementary concepts which are needed will be presented in Chapter I; here we shall formulate only the simplest properties for which analogies exist in the theory of algebraic fields, which will also be described in the introduction.

Let $(X, \mathscr{B}, \mu)$ be a probability space with the σ-algebra $\mathscr{B}$ and the countably additive probability measure μ. Let T be a measure preserving transformation from X into itself. This means that for every $A \in \mathscr{B}$, we have

$$\mu(T^{-1}A) = \mu(A).$$

We consider the powers $T^2, T^3, \ldots$ of the transformation T. Denote by $g(P)$ the indicator function of the arbitrarily chosen measurable set $A \in \mathscr{B}$; i.e.,

$$g(P) = \begin{cases} 1 & \text{if } P \in A \\ 0 & \text{if } P \notin A. \end{cases}$$

For any point $P_0 \in X$, the collection of points

$$P_0, TP_0, T^2P_0, \ldots$$

is called the (discrete) trajectory generated by the point P_0. The "relative time" which the trajectory spends in the set A is characterized by the sum

$$\frac{1}{n} \sum_{k=1}^{n} g(T^k P_0) . \tag{0.1.1}$$

In Chapter I the conditions are presented for which the sum (0.1.1) has a limit as $n \to \infty$ for almost all points $P_0 \in X$, and stronger conditions under which this limit is shown to be equal to $\mu(A)$ – the measure of the set A. Also conditions which guarantee a "mixing property" will be further discussed.

Statements in the language of ergodic theory hold true for "almost all" points P_0 of the fundamental space X and possess thus probabilistic metric character. Applications of ergodic theory exist to the metric theory of numbers. These applications lead to theorems on properties of almost all real (or complex) numbers with respect to Lebesgue measure on the line (or on the plane).

§ 2. Applications of Ergodic Concepts to the Theory of Diophantine Equations

This book is of another type than those concerned with metric number theory, since we are interested in application of the concepts of ergodic theory to number theory and mainly to the number theory of real and imaginary quadratic fields. Namely, we present here concepts useful for obtaining asymptotic results concerning solutions of certain diophantine equations which contain unboundedly increasing parameters. These diophantine equations are very closely connected with the behavior of ideal classes of certain sequences of algebraic fields. Under certain restrictions the asymptotic results are deduced for all parameter values entering into the equations, and not for almost all values (in some sense) as in metric number theory.

The diophantine equations which we investigate will contain few variables in comparison with the degrees of the variables entering into them. Therefore the results obtained on the asymptoticity of their solutions will not be attainable by the usual methods of analytic number theory for solving diophantine equations with a sufficiently large number of variables – methods developed by G. Hardy, J. Littlewood, and I. M. Vinogradov.

More details on the diophantine equations spoken of and on their connection with the theory of algebraic fields will be presented in Chapter VII.

Here we describe only one special case – the problem concerning the distribution of integral points on the sphere

(0.2.1) $$x^2+y^2+z^2=m$$

for increasing values of m.

We consider in general the sphere in n-dimensional space

(0.2.2) $$x_1^2+x_2^2+\cdots+x_n^2=m$$

for $n\geqq 1$ and we pose the problem of the asymptotic distribution of integral points on this sphere. The case $n=1$ is trivial, for $n=2$ we obtain the elementary problem of the number of fields $k(\sqrt{-1})$; here we can speak of asymptotic distributions only for special forms of the number m. Thus, if m contains even one odd power of a prime factor of the form $4k-1$, then the equation (0.2.2) for $n=2$ does not generally have a solution, as is well known. We remark, by the way, that for the case $n=2$ and for special types of numbers m, elementary ergodic considerations produce in certain cases effective results (see Chapter IX); hence the given equations (0.2.2) for $n=2$ are of some interest.

The case $n\geqq 4$ can be investigated by the usual methods of analytic number theory, where for $n\geqq 5$ no special difficulties arise and for $n=4$ difficulties appear which are overcome by the application of sufficiently deep properties of Kloosterman sums. The case $n=4$ has been investigated in detail only in recent years by A. V. Malyshev and G. Pommerenke. In Malyshev results are obtained not only valid for spheres, but also for arbitrary ellipsoids, using integral positive quadratic forms. We formulate a theorem by A. V. Malyshev for the case of the sphere (0.2.2) and $n\geqq 4$. Let $\Lambda_{n,m}$ be a convex region on the surface of the sphere which, seen from the center of the sphere, cuts out the solid angle $\lambda>0$, and let $\lambda_0=\dfrac{(2\pi)^{n/2}}{\Gamma(\frac{n}{2})}$ be the "total" solid angle. Denote by $r_m(\Lambda_{n,m})$ the number of integral points on the sphere (0.2.2) lying in the region $\Lambda_{n,m}$. Then

(0.2.3) $$r_m(\Lambda_{n,m})=\frac{\lambda}{\lambda_0}\,r_m(1+0(m^{-\gamma_n}))\,.$$

Here r_m is the total number of integral points on the sphere (0.2.2); $\gamma_n>0.02$ is a positive constant (for increasing n, γ_n approaches $\frac{1}{12}$). We note that the number r_m may be expressed in the form

(0.2.4) $$r_m=\frac{\pi^{\frac{n}{2}}m^{\frac{n}{2}-1}}{\Gamma(\frac{n}{2})}\cdot S_n(m)\,,$$

where $S_n(m)$ is a special series of arithmetic functions satisfying the conditions

$$c_1(\varepsilon)\,m^{\varepsilon}\geqq S_k(m)\geqq c_2(\varepsilon)\,m^{-\varepsilon}$$

for $\varepsilon>0$ as small as desired; $c_1(\varepsilon)$ and $c_2(\varepsilon)$ are positive constants.

Thus the integral points on the surface of the sphere (0.2.2) are asymptotically uniformly distributed, according to formula (0.2.3), for $n \geqq 4$. We remark that, for $n \geqq 4$ even, similar problems may be treated by means of elliptic modular functions (see POMMERENKE), but then the results obtained are considerably weaker than ones of the above type.

For $k=3$, that is, for the three-dimensional sphere

$$x^2 + y^2 + z^2 = m, \tag{0.2.5}$$

we can expect the same asymptotic uniform distribution of points. However, the situation here is considerably more complicated. First of all, as was established by GAUSS (see, for example, VENKOV [2]), not all spheres of the form (0.2.5) contain integral points in their surfaces. In order for a sphere to possess integral points, it is necessary and sufficient that m is not of the form $m = 4^a(8b+7)$, where a and b are non-negative integers. Therefore, if m does not have such a form, then the number r_m of integral points on the sphere (0.2.5) will be positive. Let $m = 2^{a_1} m_1$, where $a_1 \geqq 0$, $m \neq 4^a(8b+7)$, and m_1 odd. Then, as shown by K. L. SIEGEL in 1935, $r_m \to \infty$ as $m \to \infty$; more precisely, the inequalities

$$c_1(\varepsilon)\, m_1^{\frac{1}{2}+\varepsilon} \geqq r_m \geqq c_2(\varepsilon)\, m_1^{\frac{1}{2}-\varepsilon} \tag{0.2.6}$$

are valid, where $\varepsilon > 0$ is, as above, as small as desired, and $c_1(\varepsilon)$, $c_2(\varepsilon)$ are positive constants dependent on ε. K. L. SIEGEL's inequality (0.2.6) turns out to be sufficient in order to justify, under certain supplementary conditions, the asymptotic distribution of integral points on the sphere (0.2.5). Here the usual methods of analytic number theory, and also of the theory of elliptic modular functions, fail to work. We explain shortly in which manner we are led here to asymptotic statements by ergodic concepts.

Let us assume, for simplification of the affair, that m is odd and $m \neq 8b+7$, so that we may substitute m for m_1 in the equations (0.2.6). We see that on the sphere (0.2.5) there will be sufficiently many integral points and, among them, as can be shown, there will be sufficiently many primitive integral points. By primitive integral points we mean those integral points (x, y, z) for which *g.c.d.* $(x, y, z) = 1$ (*g.c.d.* designates greatest common divisor and will be omitted when the meaning is clear). The number of such points is denoted by r'_m. Then it turns out to be true that $r'_m > c\, r_m$, where c is a positive constant.

Let A be the set of all primitive points on the sphere (0.2.5). We shall establish a "flow" of integral points of A determined by a rule for which one of these integral points is carried over into another.

We fix a prime number $p \geqq 3$. We assume that m is sufficiently large in comparison to p and, in addition, that m satisfies the condition

$\left(\frac{-m}{p}\right) = +1$ (the left symbol is the Legendre symbol). Let

$$L = \begin{Vmatrix} x \\ y \\ z \end{Vmatrix}$$

be a primitive integral point on the sphere written as a column vector; this vector joins the origin (0, 0, 0) with our point (x, y, z). We consider an orthogonal matrix P whose components are rational fractions of the form a/p, where a is an integer. Among these matrices we may delete in a natural manner associated (inessentially different) ones, and then we have remaining a set of non-associated essentially different matrices $\{P\}$. The point $P \cdot L$ is the result of the action of the matrix P on L and will of course lie in the sphere (0.2.5). However, $P \cdot L$ will be integral for exactly two matrices of the set $\{P\}$. We can select in a certain manner one of these two matrices; denote it by P_L. We have

$$P_L \cdot L = L', \tag{0.2.7}$$

where L' is a new integral primitive point on the sphere. Thus equation (0.2.7) assigns to each primitive point L on the sphere a primitive point L' on the same sphere, so that (0.2.7) defines a flow on the set M of primitive points of the sphere. Starting from any point $L \in M$, we obtain by means of equations of type (0.2.7) a trajectory of the flow. Since the set M is finite, the trajectories eventually end in finite cycles. It turns out that in all cases the trajectories are themselves finite cycles. (Of course, flows on finite sets will have, generally speaking, "pre-cyclic regions", which are followed by cycles, but for the flow described above the trajectories are all of cyclic type.)

If m is sufficiently large, then it turns out that "on the average" these disjoint cycles will possess a length of the order $\log m$. Let the surface of the unit sphere

$$x^2 + y^2 + z^2 = 1 \tag{0.2.8}$$

be divided into a finite number of simply connected regions with smooth boundaries. We project these regions to the sphere (0.2.5) centrally from the origin, so that we obtain there the regions $\Lambda_1, \Lambda_2, \ldots, \Lambda_n$. Taking any cycle, we shall investigate the passing of this trajectory through the given regions $\Lambda_1, \Lambda_2, \ldots, \Lambda_m$. It happens that an "overwhelming majority" of the trajectories will be "good" in the sense that the "time of visit" of the trajectory in the region Λ_i is asymptotically proportional to the area of the projection of Λ_i onto the unit sphere (the solid angle which Λ_i covers as seen from the origin). "Poor" trajectories, for which this is not so, are relatively few in number.

The asymptotic uniform distribution of integral points on the sphere follows from these facts; points on "good" trajectories are distributed on the sphere in an asymptotically uniform manner and the number of points in "poor" trajectories is relatively small.

In this manner ergodic concepts help us to clarify asymptotic properties of solutions to the diophantine equation (0.2.5).

The described considerations can be succesfully extended to ellipsoids of sufficiently general type, i.e., to the investigation of the asymptotic geometry of solutions of the equation

$$f(x, y, z) = m, \tag{0.2.9}$$

where $f(x, y, z)$ is an integral ternary quadratic form. Here we can define a flow not only on the ellipsoid (0.2.9), but also one consisting of a transition from one ellipsoid of the form (0.2.9) to another ellipsoid of the same form, where in the place of f we have a quadratic form f' of the same Gaussian genus as f. Since the asymptoticity of representations of the number m by the genus of forms f is known, then in this manner the problem concerning representation of large numbers by ternary quadratic forms was solved, as well as the corresponding asymptoticity and "asymptotic geometry" of solutions of the equation (0.2.9). These investigation, begun by LINNIK [1] in 1940, were advanced further by A. V. MALYSHEV.

It is also possible to investigate flows on hyperboloids of one and two sheets. Here the set of primitive points may prove to be infinite, and it will be necessary to select a certain basic finite set of primitive points M and a "fundamental region" in which the flow is defined.

Problems of the stated type turn out to be closely connected with properties of ideal classes of imaginary (for ellipsoids) and real (for hyperboloids) quadratic fields and the behavior of the number of ideal classes.

With fields of higher degree and of more general form are connected systems of diophantine equations the study of which is inaccessible up to now by the usual methods of analytic number theory. Here also we succeed in constructing corresponding flows of primitive points and in preparing basically the instrumentation needed for their study. However, the corresponding construction is of considerable complexity. Therefore in this book we will only investigate the most simple results, those concerning quadratic fields, and for fields of more general types we only state the frontiers of investigation in the respective directions.

Chapter I

A Summary of Elementary Ergodic Theory and Limit Theorems of Probability Theory

§ 1. Basic Theorems of Ergodic Theory

At the end of § 1 of the introduction the probability space $(X, \mathscr{B}, \mu)$ with the σ-algebra $\mathscr{B}$ and the countably additive probability measure μ was introduced. We considered a transformation T from X into itself (not necessarily one-to-one) which was measure preserving:

$$\mu(T^{-1}A) = \mu(A) \tag{I.1.1}$$

for all $A \in \mathscr{B}$. Following the monograph of P. HALMOS [2] we shall present a few elementary theorems of ergodic theory dealing with such transformations. Denote the image of a point x under the transformation T by Tx. The first of these theorems will be the recurrence theorem, originating from A. POINCARÉ. Let $E \in \mathscr{B}$ be a measurable set. A point $x \in E$ is called recurrent (with respect to E and T) if $T^n x \in E$ for some integer $n \geqq 1$.

Recurrence Theorem: Almost all points of E (with respect to the measure μ) are recurrent.

In this connection it is easy to prove that almost all points of E return to E under the action of T infinitely many times.

We now consider measurable, absolutely integrable functions $f(x)$ mapping X into the real line $(-\infty, \infty)$; the space of all such functions will be denoted by $\mathscr{L}_1$. Then the ergodic theorem of G. BIRKHOFF holds:

Ergodic Theorem: Let $f \in \mathscr{L}_1$. Then the limit

$$\lim_{n\to\infty} \frac{1}{n} \sum_{j=0}^{n-1} f(T^j x) = f^*(x) \tag{I.1.2}$$

exists for almost all $x \in X$. Further, $f^* \in \mathscr{L}_1$ and f^* is almost invariant (i.e. $f^*(Tx) = f^*(x)$ for almost all $x \in X$. Also

$$\int_X f^*(x)\,\mu(dx) = \int_X f(x)\,\mu(dx)\,.$$

The expression on the left side of (I.1.2) is called the ergodic mean. An important special case of the ergodic theorem is obtained if one additional condition is imposed on the transformation T, the condition of ergodicity.

The set E is said to be invariant (under T) if $T^{-1}E = E$. The function f is invariant (under T) if $f(x) = f(Tx)$ for all $x \in X$, almost invariant if this equation holds true for almost all $x \in X$.

The transformation T is called ergodic (or metrically transitive) if for every invariant $E \in \mathscr{B}$ either $\mu(E)=0$ or $\mu(X-E)=0$ holds. It is easy to show that T is ergodic if and only if every almost invariant function is almost everywhere equal to a constant.

Hence we see the effect of the ergodic property on the ergodic theorem. The function $f^*(x)$ is almost invariant, thus in this case almost everywhere equal to the constant

$$\int_X f(x)\,\mu(dx),$$

that is, the mean value of $f(x)$. Thus we obtain:

Ergodic Theorem for an Ergodic Transformation T: Suppose $f \in \mathscr{L}_1$ and T ergodic. Then

$$\lim_{n\to\infty} \frac{1}{n} \sum_{j=0}^{n-1} f(T^j x) = \int_X f(x)\,\mu(dx) \tag{I.1.3}$$

holds for almost all $x \in X$.

This theorem gives an accurate interpretation of the assertion that the "space mean" and the "time mean" are near to each other in some problems in statistical physics.

An interesting special form of (I.1.3) is obtained when $f(x)$ is the indicator function of a measurable set $E \in \mathscr{B}$ (i.e., $f(x)=1$ if $x \in E$ and $f(x)=0$ if $x \in X-E$). In this case

$$\int_X f(x)\,\mu(dx) = \mu(E),$$

and we have from (I.1.3):

$$\lim_{n\to\infty} \frac{1}{n} \sum_{j=0}^{n-1} f(T^j x) = \mu(E) \tag{I.1.4}$$

for almost all $x \in X$. If we interpret the transition $T^j x \to T^{j+1} x$ in X as the movement of the points of X from time j to time $j+1$, then (I.1.4) says that the relative visiting time of the point x in the set E during the first n time periods will tend to $\mu(E)$ as n grows for almost all $x \in X$.

If E and G are two measurable subsets of X and T is ergodic, then it is not difficult to derive from (I.1.3) the relation:

$$\lim_{n\to\infty} \frac{1}{n} \sum_{j=0}^{n-1} \mu(T^j E \cap G) = \mu(E)\,\mu(G), \tag{I.1.5}$$

which is interpreted as the "mixing property": the images of the set E cover on the average a part of the given set G proportional to the size of E in X.

The ergodic concepts presented above will be useful in the sequel for the study of systems of diophantine equations; analogues of ergodic theorems will be proved and applied to obtain asymptotic solutions of such equations.

§ 2. Applications to Metric Number Theory

Applications of ergodic theorems in metric number theory are possible because both of these fields of mathematics are concerned with properties holding true for almost all points – in ergodic theory with respect to an abstract measure on an abstract space; in metric number theory with respect to Lebesgue measure on an Euclidean space. Following the monograph of M. KAC, we shall give concrete examples of some applications due to C. RYLL-NARDZEVSKI.

Consider the interval $X=(0,1)$. Any number $x \in X$ may be expanded in a continued fraction:

$$x=\cfrac{1}{a_1+\cfrac{1}{a_2+\cfrac{1}{a_3+\cdots}}},$$

where $a_1, a_2, a_3, \ldots$ are positive integers, easily expressed by the function $[x]$ (whole part of x).

For instance,

$$a_1 = a_1(x) = [1/x],$$
$$a_2 = a_2(x) = \left[\frac{1}{1/x - [1/x]}\right],$$

etc. Of course, these expressions get more complicated with increasing index of a_n. We can exhibit them, however, with the help of iterates of one and the same transformation T. We put

$$T(x) = 1/x - [1/x].$$

Then:

$$\begin{aligned} a_2(x) &= a_1(T(x)) \\ a_3(x) &= a_2(T(x)) = a_1(T^2(x)) \\ &\vdots \\ a_n(x) &= a_1(T^{n-1}(x)) \\ &\vdots \end{aligned}$$

In order to apply the ergodic theorem to the transformation T (mapping X into itself), it is necessary to define a measure on X invariant under T and to verify that T is ergodic with respect to this measure.

We shall search for an invariant measure on $(0,1)$ by looking for its density $p(x)$ with respect to Lebesgue measure; thus a function $p(x) \in \mathscr{L}_1(0,1)$ is required such that for every Lebesgue measurable set $A \subset X$ we have

$$\mu(T^{-1}A) = \int_A p(x)\,dx.$$

Instead of all measurable sets it is sufficient to consider intervals $(\alpha, \beta) \subset (0,1)$. Consider the inverse image of such an interval under the transformation T. We have

$$T^{-1}(\alpha, \beta) = \bigcup_{n=1}^{\infty} \left(\frac{1}{n+\beta}, \frac{1}{n+\alpha} \right);$$

a union of disjoint intervals, and consequently the following equation for $p(x)$ must hold:

$$\mu(T^{-1}(\alpha, \beta)) = \sum_{n=1}^{\infty} \int_{\frac{1}{n+\beta}}^{\frac{1}{n+\alpha}} p(x)\, dx .$$

If μ is to be an invariant measure, then necessarily

$$\mu(T^{-1}(\alpha, \beta)) = \mu((\alpha, \beta)) = \int_{\alpha}^{\beta} p(x)\, dx .$$

Thus we obtain the functional equation

$$\int_{\alpha}^{\beta} p(x)\, dx = \sum_{n=1}^{\infty} \int_{\frac{1}{n+\beta}}^{\frac{1}{n+\alpha}} p(x)\, dx ,$$

to hold for all α and β. It is easy to verify that the function

$$p(x) = \frac{1}{ln 2} \frac{1}{1+x}$$

satisfies this equation; further, $p(x) > 0$ and

$$\int_0^1 p(x)\, dx = 1 .$$

This solution, discovered already by GAUSS in 1812, provides us with the necessary invariant measure. We shall not have to look for other solutions; it is also easy to show that the transformation T is ergodic with respect to this measure.

Let $f(x) \in \mathscr{L}_1$ (with respect to μ), that is,

$$\frac{1}{ln 2} \int_0^1 |f(x)| \frac{dx}{1+x} < \infty .$$

Then we obtain from the ergodic theorem

$$\lim_{n \to \infty} \frac{1}{n} \sum_{j=1}^{n} f(T^j x) = \frac{1}{ln 2} \int_0^1 f(x) \frac{dx}{1+x} \tag{I.2.1}$$

for almost all $x \in X$ (with respect to μ, or equivalently, with respect to Lebesgue measure, in virtue of the equivalence of these two measures).

For the function

$$f(x) = ln\, a_1(x) = ln\, [1/x]$$

we have $f(x) \in \mathscr{L}_1$, and it follows therefore from (I.2.1) that

$$\text{(I.2.2)} \qquad \lim_{n\to\infty} \frac{1}{n} \sum_{j=1}^{n} \ln a_j(x) = c = \frac{1}{\ln 2} \int_0^1 \left[\frac{1}{x}\right] \frac{dx}{1+x},$$

and

$$\text{(I.2.3)} \qquad \lim_{n\to\infty} (a_1(x) \dots a_n(x))^{\frac{1}{n}} = e^c$$

for almost all $x \in (0, 1)$ (with respect to Lebesgue measure). This theorem, clearly of ergodic character, was first proved by A. YA. CHINTSCHIN in 1935 from a theorem of G. O. KUZMIN, which was used for direct calculations of measures of the corresponding sets; the ergodic character of the theorem was not at that time entirely realized.

§ 3. Limit Theorems of Probability Theory

In the study of flows of primitive points on surfaces, which will be developed here, Markoff chains with finite state spaces arise quite naturally (for basic results see B. V. GNEDENKO, p. 110).

We shall need some limit theorems concerning the changes of states in Markoff chains and the calculation of large deviations from the mean, especially for the number of visits to certain states as the path length grows large. In the case of homogeneous Markoff chains with finite state spaces it is possible under some "soft" conditions to reduce this problem to the theory of large deviations for sums of independent random variables. The following theorem of S. N. BERNSTEIN on probabilities of large deviations will be sufficient for further calculations.

Let $z_1, z_2, \dots, z_n$ be independent random variables with $\boldsymbol{E} z_i = 0$ and $\boldsymbol{E} z_i^2 = \beta_i > 0$ for $1 \leqq i \leqq n$.

Theorem I.3.1: *Let H be a number such that*

$$\text{(I.3.1)} \qquad |\boldsymbol{E} z_i^k| \leqq \frac{\beta_i}{2} k!\, H^{k-2}$$

for all $k \geqq 2$. Then for every $t > 0$, the probability that the inequalities

$$\text{(I.3.2)} \qquad \begin{cases} z_1 + \dots + z_n \geqq 2t\sqrt{\beta} \\ z_1 + \dots + z_n \leqq -2t\sqrt{\beta} \end{cases}$$

(where $\beta = \beta_1 + \dots + \beta_n$) are fulfilled does not exceed

$$\text{(I.3.3)} \qquad \max\left\{\exp(-t^2), \exp\left(-\frac{t\sqrt{\beta}}{2H}\right)\right\}.$$

In particular this theorem will be applied to a Bernoulli scheme with probability of success equal to $p \in (0, 1)$. Let z_i assume the values $(1-p)$ and $(-p)$ with respective probabilities p and $(1-p)$. Then $\boldsymbol{E} z_i = 0$, $\boldsymbol{E} z_i^2 = p(1-p) = \beta_i$ and $\beta = np(1-p)$. Obviously it is possible to choose a constant $H = H(p)$ such that (I.3.1) is satisfied. Now put $t = \varepsilon\sqrt{n}/2$, where $\varepsilon > 0$ is fixed. An application of Theorem I.3.1 yields:

Theorem I.3.2: *For a Bernoulli scheme of the above-mentioned type, the probability that the inequalities*

$$\left\{\begin{array}{l} z_1 + \cdots + z_n \geqq \varepsilon n \sqrt{p(1-p)} \\ z_1 + \cdots + z_n \leqq -\varepsilon n \sqrt{p(1-p)} \end{array}\right. \tag{I.3.4}$$

are fulfilled does not exceed

$$\max\left\{\exp\left(-\frac{n\varepsilon^2}{4}\right),\ \exp\left(\frac{-n\varepsilon\sqrt{p(1-p)}}{4H(p)}\right)\right\}, \tag{I.3.5}$$

where $H(p)$ is a constant chosen to satisfy (I.3.1).

The connection between this limit theorem for the estimation of large deviations and quaternions or matrices is established in LINNIK [4].

Chapter II

A Summary of the Arithmetic of Quaternions and Matrices

§ 1. Arithmetic of Quaternions

Diophantine equations of the type we shall study are closely connected with the invariants of matrices M under similarity transformations; i.e., transformations of the form $M \to M' = SMS^{-1}$ and the coefficients of the characteristic equations of such matrices.

For the investigation of the equation

$$x^2 + y^2 + z^2 = m, \tag{II.1.1}$$

or, equivalently, the distribution of integral points on the sphere, it is convenient to make use of the algebra of ordinary quaternions $\{1, i, j, k\}$; the properties of this algebra are assumed to be known. We recall that this algebra is isomorphic to an algebra of 4×4 matrices given by the representation

$$1 \to \begin{pmatrix} 1 & 0 & 0 & 0 \\ 0 & 1 & 0 & 0 \\ 0 & 0 & 1 & 0 \\ 0 & 0 & 0 & 1 \end{pmatrix} \qquad i \to \begin{pmatrix} 0 & -1 & 0 & 0 \\ 1 & 0 & 0 & 0 \\ 0 & 0 & 0 & -1 \\ 0 & 0 & 1 & 0 \end{pmatrix}$$

$$j \to \begin{pmatrix} 0 & 0 & -1 & 0 \\ 0 & 0 & 0 & 1 \\ 1 & 0 & 0 & 0 \\ 0 & -1 & 0 & 0 \end{pmatrix} \qquad k \to \begin{pmatrix} 0 & 0 & 0 & -1 \\ 0 & 0 & -1 & 0 \\ 0 & 1 & 0 & 0 \\ 1 & 0 & 0 & 0 \end{pmatrix}.$$

For information concerning the arithmetic of quaternions the reader is referred to B. A. VENKOV [3], LINNIK [2], and MALYSHEV. Here we enumerate briefly the results which we shall need; for proofs see LINNIK [2] and MALYSHEV.

Consider the quaternion algebra $\{\alpha + \beta i + \gamma j + \delta k\}$ with real coefficients. Denote by O_0 the ring of integral quaternions – integral meaning those quaternions whose coefficients α, β, γ, δ are either all integers or all halves of odd integers. In the sequel quaternions of the first type will be called properly integral (p.i.), those of the second type improperly integral (i.i.).

Integral quaternions with norm equal to one are called units of the ring O_0. The 24 units are:

$$\pm 1,\ \pm i,\ \pm j,\ \pm k,\ \frac{\pm 1 \pm i \pm j \pm k}{2}.$$

Elements of the ring O_0 will be designated by Latin capitals. The norm of A is represented by $Norm(A)$ and is equal to the sum of squares of the coefficients of A. We have

$$Norm(A) = A\bar{A} = \bar{A}A,$$

where $\bar{A}$ is the conjugate of A.

If $A = BC$, then we say that A is left divisible by B, right divisible by C.

An analogy to the Euclidean algorithm is valid for quaternions. To each pair A and B of internal quaternions with $B \neq 0$ there exists $C \in O_0$ such that

$$A = BC + D$$

and

$$Norm(D) < Norm(B)$$

(and analogously for right division by B).

The quaternion D_1 is called the left greatest common divisor of A and B if $A = D_1 A_1$, $B = D_1 B_1$, and

$$D_1 = AX_1 + BX_2; \quad X_1, X_2 \in O_0$$

(and analogously for right g.c.d.). Each pair of integral quaternions A, B possesses a left g.c.d. and a right g.c.d.

The quaternion A is called primitive if it is not divisible by any integer greater than 1.

Two quaternions whose left g.c.d. (right g.c.d.) is a unit are said to be left (right) relatively prime. If two quaternions are left and right relatively prime, then they are said to be relatively prime. If $Norm(A)$ is a prime number, then A is called a prime quaternion.

Each primitive quaternion has a norm which is not divisible by 4. If its norm is written in the form

$$Norm(A) = 2^{\alpha_0} p_1^{\alpha_1} \dots p_l^{\alpha_l}; \quad \alpha_0 \geqq 0; \quad \alpha_1, \dots, \alpha_l > 0,$$

where $p_1, \dots, p_l$ are odd primes, then α_0 is equal to 0 or 1 and A has the form

(II.1.2) $$A = (1+i)^{\alpha_0} P_{11} \dots P_{1\alpha_1} P_{21} \dots P_{2\alpha_2} \dots P_{l1} \dots P_{l\alpha_l},$$

where the P_{ij} are prime quaternions with norm p_i.

The representation (II.1.2) is unique in the following sense: for any other representation of A with the same $p_1, \dots, p_l$ and $\alpha_0, \dots, \alpha_l$, let us say

$$A = (1+i)^{\alpha_0} P'_{11} \dots P'_{1\alpha_1} \dots P'_{l1} \dots P'_{l\alpha_l},$$

we have

$$P'_{11} = P_{11}\varepsilon, \quad P'_{12} = \bar{\varepsilon} P_{12} \varepsilon', \quad P'_{13} = \bar{\varepsilon}' P_{13} \varepsilon'', \quad \dots$$

etc., where $\varepsilon, \varepsilon', \varepsilon'', \dots$ are suitable units.

Quaternions differing only by a unit factor on the left (right) are said to be left (right) associated.

Let $U(n)$ be the number of integral quaternions with norm n. Then $U(n) = 24\bar{\sigma}(n)$, where $\bar{\sigma}(n)$ is the sum of all odd divisors of n. In particular, $U(2^\lambda) = 24$ for $\lambda \geqq 0$. If n is equal to a prime p, then $U(p) = 24(p+1)$.

Let A be an arbitrary quaternion with odd norm. If S is a quaternion, then the set of all quaternions of the form $S+AX$, where X runs over all integral quaternions, is called a left residue class ($mod\,A$), and the set of all $S+XA$ is called a right residue class ($mod\,A$). For each quaternion $A \in O_0$ with odd norm we have:

The number of left (right) residue classes ($mod\,A$) is equal to $(Norm(A))^2$.

The collection of quaternions of the form $\alpha S+AX$, where α runs through all integers and X through all integral quaternions, is called a left ray ($mod\,A$). Right rays ($mod\,A$) are analogously defined. Rays are in particular modules over the ring of integers.

If A is a prime quaternion with $Norm(A)=p \geqq 3$, then the number of right (left) rays ($mod\,A$) is equal to $p+1$.

If $S=1$, then the ray $\alpha+AX$ is called the principal left ray ($mod\,A$); the principal right ray ($mod\,A$) is analogously defined. Principal rays are rings. In the sequel the following theorem, found in LINNIK [2] and in a more general form in MALYSHEV, is important.

Theorem II.1.1: *Let M und M' be two primitive quaternions with the same odd norm m. Let S be any quaternion with the properties*

$$left\ g.c.d.\ (S, M')=1$$

and

$$SM = M'S'. \tag{II.1.3}$$

Such a quaternion S always exists, and the set of all quaternions Q with the property

$$QM = M'Q' \tag{II.1.4}$$

is identical with the left ray ($mod\,M'$) $\alpha S+M'X$.

A similar assertion holds true for division by M and M' on the other sides: If M and M' are primitive with odd norm m and $MS=S'M'$, *right g.c.d.* $(S, M')=1$, then the set of all quaternions satisfying $MQ=Q'M'$ is identical with the ray $\alpha S+XM'$. Such a quaternion S always exists. In particular, if $M=M'$ and $S=1$, then the set of all quaternions Q for which

$$QM = MQ' \tag{II.1.5}$$

coincides with the principal ray $\alpha+MX$. The analogous statement holds true for quaternions Q satisfying

$$MQ = Q'M. \tag{II.1.6}$$

An important case in the sequel for the pair M and M' turns out to be the case $M'=\bar{M}$; we consider the equation

$$QM = \bar{M}Q'. \tag{II.1.7}$$

By theorem (II.1.1) these quaternions Q form a left ray ($mod\,\bar{M}$), which will be called a versored ray. The following theorem holds (see LINNIK [2] and MALYSHEV):

Theorem II.1.2: *The left versored ray* ($mod\,\bar{M}$) *is made up of exactly those quaternions for which*

$$2\mathcal{R}(QM) \equiv 0\,(mod\,m) \tag{II.1.8}$$

(where $\mathcal{R}$ *designates the real part of a quaternion).*

The right versored ray ($mod\,\bar{M}$) is treated analogously. Further special properties of quaternions in connection with rotations of spheres will be given at the beginning of Chapter III.

§ 2. Arithmetic of 2×2 Matrices

The arithmetic of 2×2 matrices is similar to that of quaternions; the difference turns out to be the presence of divisors of zero and of infinitely many units. We collect here a few basic facts, referring the reader to Jacobson and Davenport for details.

Let us consider the ring of all matrices

$$\begin{pmatrix} \alpha & \beta \\ \gamma & \delta \end{pmatrix}$$

over the real numbers, in which the matrix

$$\begin{pmatrix} 1 & 0 \\ 0 & 1 \end{pmatrix}$$

will be identified with the number 1. Each matrix A may be represented in the form

$$A = l + L,$$

putting

$$l = \frac{1}{2} Sp(A) = \frac{\alpha+\delta}{2},$$

$$L = A - l.$$

Then

$$L = \begin{pmatrix} \frac{\alpha-\delta}{2} & \beta \\ \gamma & -\frac{\alpha-\delta}{2} \end{pmatrix}$$

is a matrix with zero trace. Let us agree to call l the scalar part of A and L the vector part of A. If $l=0$, then $A=L$ will be called a vector matrix. For a vector matrix

$$L = \begin{pmatrix} \mu & \beta \\ \gamma & -\mu \end{pmatrix}$$

we have

$$L^2 = \mu^2 + \beta\gamma$$

so that $L^2 = -D$. (D is the determinant of L.)

If

$$A=\begin{pmatrix}\alpha & \beta\\ \gamma & \delta\end{pmatrix}=l+L,$$

then the matrix

$$\bar{A}=l-L=\begin{pmatrix}\delta & -\beta\\ -\gamma & \alpha\end{pmatrix}$$

will be called the conjugate of A. We have

$$A\bar{A}=\bar{A}A=\alpha\delta-\beta\gamma=det(A).$$

Matrices with nonzero determinants will be called, as is usual, regular. The inverse of a regular matrix A is

$$A^{-1}=\frac{\bar{A}}{det(A)}.$$

Consider now the vector matrix

(II.2.1) $$L=\begin{pmatrix}b & -a\\ c & -b\end{pmatrix}.$$

The similarity transformation

(II.2.2) $$L'=ALA^{-1}$$

of the vector matrix L with the regular matrix

$$A=\begin{pmatrix}\alpha & \beta\\ \gamma & \delta\end{pmatrix}$$

results in a new vector matrix L' with $L'^2=-D$.

Lemma II.2.1 gives more detailed information about this transformation.

Lemma II.2.1: *Let*

$$A=\begin{pmatrix}\alpha & \beta\\ \gamma & \delta\end{pmatrix}$$

be regular with $det(A)=u>0$. *Then*

(II.2.3) $$A\begin{pmatrix}b & -a\\ c & -b\end{pmatrix}A^{-1}=\begin{pmatrix}b' & -a'\\ c' & -b'\end{pmatrix},$$

in which a' is positive if a is positive, and the quadratic form

$$\varphi'(x,y)=a'x^2+2b'xy+c'y^2$$

is obtained from $\varphi(x,y)$ by the substitution

$$\frac{1}{\sqrt{u}}\begin{pmatrix}\alpha & \gamma\\ \beta & \delta\end{pmatrix}=\frac{1}{\sqrt{u}}A^T.$$

In other words,

(II.2.4) $$\varphi'(x,y)=\varphi(x,y)S;\quad S=\frac{1}{\sqrt{u}}A^T.$$

This lemma is verified by an immediate calculation.

In particular, if the matrix A is integral and unimodular, then $\varphi'(x, y) = \varphi(x, y) A^T$ is equivalent to $\varphi(x, y)$. We obviously obtain all forms equivalent to the given form $\varphi(x, y)$ if in the transformation (II.2.4) we let A run through all unimodular matrices.

The algebra of matrices

$$A = \begin{pmatrix} \alpha & \beta \\ \gamma & \delta \end{pmatrix}$$

possesses the inconvenient property that its "multiplicative" norm $det(A) = \alpha\delta - \beta\gamma$ is not "additive"; i.e., it is not the sum of functions each depending on one argument. Therefore it will be sometimes necessary to pass to a new algebra $\mathscr{E}$ (hermitions; see MALYSHEV) with a norm having the "additive" form. Here we set

(II.2.5) $$1 = \begin{pmatrix} 1 & 0 \\ 0 & 1 \end{pmatrix}; \quad i_1 = \begin{pmatrix} 0 & -1 \\ 1 & 0 \end{pmatrix}; \quad i_2 = \begin{pmatrix} 1 & 0 \\ 0 & -1 \end{pmatrix}; \quad i_3 = \begin{pmatrix} 0 & -1 \\ -1 & 0 \end{pmatrix}.$$

Then we obtain

(II.2.6) $$\begin{cases} i_1^2 = -1, \; i_2^2 = i_3^2 = 1, \; i_\alpha i_\beta = -i_\beta i_\alpha \; (\alpha \neq \beta), \\ i_1 i_2 = -i_3, \; i_2 i_3 = i_1, \; i_3 i_1 = -i_2, \end{cases}$$

where the hermitions and matrices are related by

(II.2.7) $$A = \xi + \mu_1 i_1 + \mu_2 i_2 + \mu_3 i_3 = \begin{pmatrix} \xi + \mu_2 & -(\mu_1 + \mu_3) \\ (\mu_1 - \mu_3) & \xi - \mu_2 \end{pmatrix}$$

and

(II.2.8) $$A_1 = \begin{pmatrix} \alpha & \beta \\ \gamma & \delta \end{pmatrix} = \frac{\alpha + \delta}{2} + \frac{\gamma - \beta}{2} i_1 + \frac{\alpha - \delta}{2} i_2 + \frac{-(\beta + \gamma)}{2} i_3 .$$

For $A = \xi + \mu_1 i_1 + \mu_2 i_2 + \mu_3 i_3$ we set

$$\bar{A} = \xi - \mu_1 i_1 - \mu_2 i_2 - \mu_3 i_3 ,$$
$$Norm(A) = A\bar{A} .$$

If $\xi = 0$, we shall call A a vector; in this case $A^2 = -Norm(A)$. The matrix

$$L = \begin{pmatrix} b & -a \\ c & -b \end{pmatrix}$$

is represented by the hermition

$$L = \frac{a + c}{2} i_1 + b i_2 + \frac{a - c}{2} i_3 .$$

We shall be interested in integral vector matrices L as in (II.2.1) and mappings of L into integral L' of the form (II.2.2). We shall need some facts from the theory of integral matrices and their corresponding hermitions in $\mathscr{E}$. The matrix

$$A = \begin{pmatrix} \alpha & \beta \\ \gamma & \delta \end{pmatrix}$$

will be called integral if all of its coefficients are integers. Hence it is necessary to call the hermition

$$X = \xi + x_1 i_1 + x_2 i_2 + x_3 i_3$$

integral, if ξ, x_1, x_2, x_3 are integral (properly integral) or if X has the form

$$X = X_0 + m_1 e_1 + m_2 e_2 ,$$

where X_0 is properly integral, m_1, m_2 are integers, and

$$e_1 = \pm 1/2 \pm 1/2\, i_2 , \quad e_2 = \pm 1/2\, i_1 \pm 1/2\, i_3 .$$

If X is integral and possesses one or more non-integral components, then we shall call X improperly integral.

The divisors of zero in the ring of integral matrices (or hermitions) are singular matrices, with $det(A) = 0$. However, the Euclidean algorithm is valid: if A and B belong the ring of integral matrices V, then there exist D, X, Y in the ring V such that

$$A = DA_1 ,$$
$$B = DA_2 ,$$

and

$$AX + BY = D ,$$

as well as D', X', Y' in V such that

$$A = A_1' D' ,$$
$$B = B_1' D' ,$$

and

$$X'A + Y'B = D' .$$

This can be shown for the ring $\mathscr{E}$ by using the same methods as are used for quaternions. (For the Euclidean algorithm in the ring of integral $n \times n$ matrices for arbitrary n see KUBILIUS.)

The Euclidean algorithm permits the development of the arithmetic of matrices parallel to the lines followed in § 1 of this chapter for quaternions. The matrix A is called primitive if not all of its coefficients are divisible by an integer greater than 1. If A is primitive and

$$det(A) = 2^{\alpha_0} p_1^{\alpha_1} \dots p_l^{\alpha_l} ,$$

then

$$A = U_1 U_2 \dots U_{\alpha_0} P_{11} \dots P_{1\alpha_1} P_{21} \dots P_{2\alpha_2} \dots P_{l1} \dots P_{l\alpha_l} ,$$

where $det(U_j) = 2$, $det(P_{ju}) = p_j$.

If

$$A = U_1' \dots U_{\alpha_0}' P_{11}' \dots P_{l\alpha_l}' = U_1'' \dots U_{\alpha_0}'' P_{11}'' \dots P_{l\alpha_l}''$$

are two such representations, then

$$U_1'' = U_1' \varepsilon , \quad U_2'' = \bar{\varepsilon} U_2' \varepsilon' , \quad U_3'' = \bar{\varepsilon}' U_3' \varepsilon'' , \quad etc.,$$

where ε, ε', ε'', ... are suitably chosen unimodular matrices. (We express this by saying that the unique factorization theorem correct up to units is valid for integral matrices.)

The theory of "rays" is also developed as in § 1 of this chapter.

If M and M' are two primitive matrices such that $det(M) = det(M') = m > 0$; m odd, then there exists a primitive matrix Q with $det(Q) = q > 0$, $(q, m) = 1$, and $QM = M'Q'$. The analogous theorem holds true for right multiplication.

The set of matrices $R \in V$ with $det(R) \lessgtr 0$ and $RM = M'R'$ form a left ray ($mod\, M'$). They have the form

$$R = \alpha Q + M'X,$$

where α is an integer, Q one of the matrices indicated above. The concept right ray ($mod\, M'$) is defined analogously. For prime matrices P with $det(P) \geqq 3$ there are $p+1$ left rays ($mod\, P$) and $p+1$ right rays ($mod\, P$). This results in the following theorem (see LINNIK [2]):

The number $U(p)$ of prime matrices of norm $p \geqq 3$ which are not left (right) associated is given by

$$U(p) = p + 1.$$

An elementary calculation shows the truth of this statement and also yields for $p = 2$: $U(2) = 3$. For 2×2 matrices this theorem is easily proved without using the theory of rays.

The concepts of principal rays and versored rays (as in § 1 of this chapter) turn out to be essential here also. We formulate them only for left multiplication; the analogous facts for right multiplication are true also.

The principal left ray ($mod\, M$) translates M into itself. It has the form $\alpha + MX$. The set of matrices belonging to this ray form a ring with interesting properties. The left versored ray ($mod\, M$) transforms the matrix

$$M = \begin{pmatrix} \alpha & \beta \\ \gamma & \delta \end{pmatrix}$$

into its conjugate

$$\bar{M} = \begin{pmatrix} \delta & -\beta \\ -\gamma & \alpha \end{pmatrix}.$$

It consists of exactly those matrices Q for which

$$Sp(QM) \equiv 0 \ (mod\, m);$$

i.e., twice the "real part" is divisible by m.

It is necessary to mention a further fact of the type developed in § 1 of this chapter, namely the theorem about the number $U(n)$ of primitive left (right) non-associated matrices R with odd $det(R) = n > 0$. If

$$n = p_1^{\alpha_1} \dots p_l^{\alpha_l}; \quad \alpha_i > 0,$$

then

$$U(n) = (p_1 + 1)\, p_1^{\alpha_1 - 1} \dots (p_l + 1)\, p_l^{\alpha_l - 1}.$$

We also need the following lemma (see Theorem III.1.2):

Let A and B be primitive matrices with AB non-primitive and divisible by the integer $d > 1$. Then there exists an integral matrix D with $det(D) = d$ such that

$$A = A_1 D, \quad B = \bar{D} B_1 .$$

§ 3. Arithmetic of $n \times n$ Matrices

The fundamental concepts of the arithmetic of integral matrices

$$A = \begin{pmatrix} a_{11} & \dots & a_{1n} \\ \vdots & & \vdots \\ a_{n1} & \dots & a_{nn} \end{pmatrix}$$

i.e., matrices with integral components a_{ij}, can be found in the monographs JACOBSON and DAVENPORT.

The quantity $|det(A)|$ plays the role of the norm; thus in this case we also have divisors of zero. The Euclidean algorithm holds true, and the unique factorization theorem correct up to units (integral matrices with determinant ± 1), as described in § 2 for 2×2 matrices, is also valid here. A matrix is said to be primitive if it is not divisible by any integer greater than 1.

If M and M' are two primitive matrices with $det(M) = det(M') = m > 0$, m odd, then there exists a primitive matrix Q such that $det(Q) = q > 0$, $(q, m) = 1$, and

$$QM = M'Q' .$$

In addition, the following theorem is also valid, and we shall formulate it again in Chapter VII (Theorem VII.2.2).

Let L and Q be integral matrices with det(Q) square-free. Then the matrix

$$L' = Q^{-1} L Q$$

is integral if and only if there exists an integer l, a unit matrix E, and an integral matrix U such that

$$lE + L = QU . \qquad \text{(II.3.1)}$$

This theorem plays a role in the theory of "rotations" of matrices; we shall consider this subject in greater detail in Chapter VII, § 3.

If A is a given matrix, then its conjugate will be denoted by $\bar{A}$: $A\bar{A} = det(A)$. The transpose of A will be denoted by A^T.

Chapter III

Rotations of the Sphere, Binary Quadratic Forms, and Quaternions

§ 1. Supplementary Arithmetic Information

Let m be an integer such that $m > 3$, $m \equiv 1, 2 \pmod 4$, or $m \equiv 3 \pmod 8$. We can write the equation of the sphere

$$x^2 + y^2 + z^2 = m \tag{III.1.1}$$

with the help of quaternions. We introduce the properly integral quaternion $L = xi + yj + zk$, and replace equation (III.1.1) by the quaternion equation

$$L^2 = -m\,. \tag{III.1.2}$$

A quaternion L with $\mathscr{R}(L) = 0$ will be called a (three-dimensional) vector. Primitive solutions of (III.1.1) are said to be primitive vectors. The existence and number of solutions of (III.1.1) are described by a theorem of Gauss (see for instance B. A. Venkov [2]). In this vein we consider binary quadratic forms

$$(a, b, c) = ax^2 + 2bxy + cy^2$$

with determinant $(-m)$: $b^2 - ac = -m$.

We shall call the quadratic form properly primitive if $(a, 2b, c) = 1$, and improperly primitive if $(a, 2b, c) = 2$. The number $H_0(m)$ of solutions of (III.1.1) is given by Gauss' theorem:

Theorem III.1.1:

$$H_0(m) = \begin{cases} 12\,h(-m) & \text{if} \quad m \equiv 1, 2 \pmod 4 \\ 24\,h'(-m) = 8\,h(-m) & \text{if} \quad m \equiv 3 \pmod 8\,, \end{cases}$$

where $h(-m)$ is the class number of properly primitive forms with determinant $(-m)$, $h'(-m)$ the class number of improperly primitive forms with determinant $(-m)$.

Moreover, the problem of determining all integral points on the sphere (III.1.1) is easily reduced to that of determining the primitive points on spheres with radii of the form $\sqrt{m}/\delta$, where δ^2 divides m. In studying the distribution of primitive integral points on the sphere (III.1.1) it is convenient to use rotations of the sphere carrying primitive integral points into the same. These rotations were first investigated by B. A. Venkov [1], who deduced with their help Theorem III.1.1, showing the existence of at least one solution. A detailed exposition of the purely arithmetical aspects of the theory of rotations of this type is found in the indicated works of B. A. VENKOV, and in a simpler form with some modifications in LINNIK [2] and the monograph of A. V. MALYSHEV. We present, without proofs, some facts which will be useful to us, referring the reader to the cited works above. Setting as a goal the consideration of the ergodic aspects of the described phenomena, we shall not in detail dwell on the corresponding algebraic aspects. However, for some theorems of special character the proofs will be given.

Let L and L' be primitive vectors with norm m, that is, primitive solutions of the equation (III.1.2). Then there exist integral primitive quaternions A, C and an integer b such that

(III.1.3) $$b+L=AC,\ CLC^{-1}=\bar{A}L\bar{A}^{-1}=L',\ b+L'=CA\,.$$

The ordered pair (L, L') will be called a rotation (from L to L'). The rotation (L, L') will be compared with the positive binary form (a, b, c) with determinant $(-m)$:

(III.1.4) $$(a, b, c)=ax^2+2bxy+cy^2=\mathrm{Norm}(\bar{A}x+Cy)\,,$$

so that we have:

$$a=\mathrm{Norm}(A),\quad c=\mathrm{Norm}(C),\quad b=\mathscr{R}(AC),\quad b^2-ac=-m\,.$$

We shall say that the form (a, b, c) governs the rotation (L, L'). The collection of such forms for given L and L' forms a class of forms with determinant $(-m)$. If

$$m\equiv 1, 2\ (\mathrm{mod}\,4)\,,$$

then this class is properly primitive; if

$$m\equiv 3\ (\mathrm{mod}\,8)\,,$$

then one of the classes belonging to the rotations (L, L') and $(L, (1+i)^{-1}L'(1+i))$ is properly primitive and the other is improperly primitive.

Let L be a primitive vector with odd norm $m>3$; for $m\equiv 1\ (\mathrm{mod}\,4)$ let a class of properly primitive forms with determinant $(-m)$ be given, or for $m\equiv 3\ (\mathrm{mod}\,8)$ a class of improperly primitive forms with the same determinant. Then exactly 12 different primitive vectors L' with norm m can be found for which the rotation (L, L') is governed by this class. All of these rotations have the form $(L, \varepsilon L'\varepsilon^{-1})$, where (L, L') is one such rotation and ε runs through all unit quaternions not connected by the equation $\varepsilon'=\pm\varepsilon$. Such a set of similar rotations $(L, \varepsilon L'\varepsilon^{-1})$ will be called a sheaf of rotations.

The following useful property is valid: let L, L' be two primitive vectors of norm m and r a given fixed odd integer. Then it is possible to find an integral primitive quaternion C such that

$$CLC^{-1}=L',\quad b+L=AC,\quad (\mathrm{Norm}(C), r)=1\,.$$

If $m \equiv 1, 2 \pmod 4$, then it is possible to choose C such that even $(\mathrm{Norm}(C), 2r) = 1$. The form $(\mathrm{Norm}(A), b, \mathrm{Norm}(C))$ governs the rotation (L, L').

We now turn to arithmetic theorems of a more special nature given in LINNIK [10] and MALYSHEV.

Theorem III.1.2: *Let A and B be primitive quaternions with AB being divisible by the integer d. Then there exists an integral quaternion D with norm d such that*

(III.1.5) $$A = A_1 D, \quad B = \bar{D} B_1 .$$

Proof: Let $AB = dC$. The assertion is easily reduced to the case where $\mathrm{Norm}(A)$ and $\mathrm{Norm}(B)$ are powers of the same prime q. Then d and $\mathrm{Norm}(C)$ are also powers of q.

The proof proceeds now by induction on the number of prime divisors Q of the quaternion A.

1. $A = Q_1$. Then $Q_1 B = dC$, $\bar{Q}_1 Q_1 B = d\bar{Q}_1 C$, $qB = d\bar{Q}_1 C$; because B is primitive, we have $q = d$, $A = 1 \cdot Q_1$, $B = \bar{Q}_1 C$, $\mathrm{Norm}(Q_1) = q = d$.

2. Let the theorem be true for quaternions A with less than n prime divisors. Then we shall prove it for $A = Q_1 \dots Q_n$. We have $AB = dC$, $Q_1 A' B = dC$, where

$$A' = Q_2 \dots Q_n,\ qA'B = d\bar{Q}_1 C,\ A'B = d/q\, \bar{Q}_1 C .$$

Therefore, two cases are possible:

a. $\bar{Q}_1 C = C'$ primitive, or

b. $\bar{Q}_1 C$ not primitive; then by virtue of 1. we have $C = Q_1 C'$. In case a. by the induction hypothesis:

$$A'B = d/q\, \bar{Q}_1 C,\ A' = A_1' D',\ B = \overline{D'} B',\ \mathrm{Norm}(D') = d/q ,$$
$$A_1' B_1' = C' = \bar{Q}_1 C .$$

Since A is primitive, the only possibility is: $A_1' = 1$, $B' = \bar{Q}_1 C$, and then we set $D = Q_1 D'$. In case b. we have $A'B = dC$, $A' = A_1' D$, $B = \bar{D} B_1$, and the theorem is proved.

The following important approximation theorem comes from the classical theory of binary quadratic forms.

Theorem III.1.3: *Let $\varphi(x, y)$ be a fixed binary quadratic form with determinant d and factor δ. Then the number of representations of $\varphi(x, y)$ as a sum of three squares:*

(III.1.6) $$\varphi(x, y) = (a_1 x + b_1 y)^2 + (a_2 x + b_2 y)^2 + (a_3 x + b_3 y)^2$$

under the conditions

(III.1.7) $$\begin{cases} \text{a.} & (a_1, a_2, a_3) = 1 , \\ \text{b.} & \delta \,\Big|\, \left(\begin{vmatrix} a_2 & a_3 \\ b_2 & b_3 \end{vmatrix}, \begin{vmatrix} a_3 & a_1 \\ b_3 & b_1 \end{vmatrix}, \begin{vmatrix} a_1 & a_2 \\ b_1 & b_2 \end{vmatrix} \right) , \end{cases}$$

does not exceed

$$\text{(III.1.8)} \qquad c(\varepsilon)\, d^{\varepsilon}\, [(d/\delta^2,\ \delta)]^{\frac{1}{2}}$$

for arbitrary $\varepsilon > 0$.

Proof: We shall employ the method of GAUSS for definite improper representations of forms $\varphi(x, y)$ as in (III.1.6). (For the terminology and substance of the Gaussian theory see B. A. VENKOV [2].)

1. First of all we see that it is sufficient to estimate the number of representations belonging to the given divisor

$$e = \left(\begin{vmatrix} a_2 & a_3 \\ b_2 & b_3 \end{vmatrix}, \begin{vmatrix} a_3 & a_1 \\ b_3 & b_1 \end{vmatrix}, \begin{vmatrix} a_1 & a_2 \\ b_1 & b_2 \end{vmatrix} \right).$$

After that, we establish that $e^2 \mid d$ and that the number of divisors of d does not exceed $c'(\varepsilon)\, d^{\varepsilon}$.

2. Since $(a_1, a_2, a_3) = 1$, then, according to the above-mentioned theory, the number of required representations does not exceed the number of λ among $0, 1, \ldots, e-1$ for which the form φS_λ, where

$$S_\lambda = \begin{bmatrix} 1 & -\lambda/e \\ 0 & 1/e \end{bmatrix},$$

is integral, multiplied by the number of primitive representations (III.1.6).

3. We set

$$\varphi(x, y) = \delta(\alpha x^2 + 2\beta x y + \gamma y^2),\ \alpha\gamma - \beta^2 = d/\delta^2\,.$$

With no loss of generality we may assume $(e, \alpha) = 1$ or $(e, \alpha) = 2$; otherwise the form $\varphi(x, y)$ can be replaced by an equivalent form for which this is true.

The coefficients of the transformed form

$$\varphi \cdot \begin{bmatrix} 1 & -\lambda/e \\ 0 & 1/e \end{bmatrix}$$

must be integers. Writing $e = e_1 \delta$, we find that λ has to satisfy the congruences

$$\text{(III.1.9)} \qquad \begin{cases} \alpha\lambda - \beta \equiv 0 \pmod{e_1} \\ (\alpha\alpha - \beta)^2 \equiv -d/\delta^2 \pmod{e_1^2 \delta}\,, \end{cases}$$

where $0 \leqq \lambda < e = e_1 \delta$.

We estimate the number of such λ from above. Let $\alpha\lambda - \beta = e_1 t$, where t is an integer. Then

$$t^2 \equiv \frac{d}{\delta^2 e_1^2} \pmod{\delta}$$

and the number of distinct $t \pmod{d}$ does not exceed

$$c'(\varepsilon)\delta^{\varepsilon} \left[\left(\frac{d}{\delta^2 e_1^2}, \delta \right) \right]^{\frac{1}{2}} \leqq c'(\varepsilon) d^{\varepsilon} \left[\left(\frac{d}{\delta^2}, \delta \right) \right]^{\frac{1}{2}}.$$

If $t' \equiv t'' \pmod{\delta}$, then, determining λ' and λ'' from the equations $\alpha\lambda' - \beta = e_1 t', \alpha\lambda'' - \beta = e_1 t''$, we get:

$$\alpha\lambda' \equiv \alpha\lambda'' \pmod{e_1\delta}, \quad \lambda' \equiv \lambda'' \pmod{e_1\delta}, \quad \lambda' = \lambda'',$$

since $0 \leqq \lambda' < e_1\delta$, $0 \leqq \lambda'' < e_1\delta$. Hence the number of different λ satisfying (III.1.9) does not exceed

$$c'(\varepsilon) d^{\varepsilon} [(d/\delta^2, \delta)]^{\frac{1}{2}},$$

which is the estimate (III.1.8).

Another useful approximation theorem follows more easily:

Theorem III.1.4: *The class number of positive binary quadratic forms φ with determinant d and given minimum does not exceed*

$$c(\varepsilon) d^{\varepsilon} [(\min\varphi, d)]^{\frac{1}{2}} \tag{III.1.10}$$

for arbitrary $\varepsilon > 0$.

Proof: Let (a, b, c) be a reduced binary form with determinant $b^2 - ac = -d$ and $\min\varphi = c$. Then the number of such forms, or equivalently, the class number, does not exceed the number of solutions to the congruence

$$b^2 \equiv -d \pmod{c}.$$

The number of solutions to this expression is estimated by the quantity (III.1.10).

Another special arithmetic theorem which will be proved here is highly valuable for the study of ergodic properties of rotations. It concerns the set-up of "consistent" binary quadratic forms in the sense of Gauss-Dirichlet (see B. A. VENKOV [2]).

Let the properly or improperly primitive form (a, b, c) govern the rotation (L, L_1) through the corresponding equations

$$b + L = AC, \quad CLC^{-1} = L_1, \quad \mathrm{Norm}(A) = a, \quad \mathrm{Norm}(C) = c. \tag{III.1.11}$$

Further, let the rotation (L_1, L_2) be governed by the form (a_1, b_1, c_1) through the equations

$$b_1 + L_1 = A_1 C_1, \quad C_1 L_1 C_1^{-1} = L_2, \quad \mathrm{Norm}(A_1) = a_1, \quad \mathrm{Norm}(C_1) = c_1. \tag{III.1.12}$$

Theorem III.1.5: *Under the above conditions the rotation (L, L_2) is governed by the class of forms (a_2, b_2, c_2) which is the Gaussian composition of the classes (a, b, c) and (a_1, b_1, c_1).*

Proof: For fixed (a, b, c) we choose a representative (a_1, b_1, c_1) such that a_1 is odd and $(a, a_1) = 1$. This is possible because (a_1, b_1, c_1) is a properly primitive form. Then we have

$$(AA_1)^{-1} L (AA_1) = L_2,$$

or

$$LAA_1 = AA_1 L_2.$$

Observe that AA_1 is a primitive quaternion, since $(a, a_1)=1$ and A and A_1 are primitive. We see further that L belongs to the left principal ray mod AA_1. It follows that there exists an integer μ with

(III.1.13) $$\mu+L=AA_1V, \quad V \text{ integral}.$$

Consulting (III.1.11), we obtain

$$\mu-b=AW, \quad W \text{ integral},$$

from which follows (see Chapter II)

(*) $$\mu\equiv b \pmod{a}.$$

From (III.1.13) we extract

$$\mu+L_1=A_1W_1, \quad W_1=VA,$$

and from (III.1.12) follows

$$\mu-b_1=A_1\mathrm{W}_2,$$

hence

(**) $$\mu\equiv b_1 \pmod{a}.$$

Looking at (III.1.13) we see that the form $\psi=(aa_1, \mu, \mathrm{Norm}(V))$ governs the rotation (L, L_2). Further, $(a, a_1, b+b_1)=1$, so that the forms (a, b, c) and (a_1, b_1, c_1) are consistent. In virtue of the conditions (*) and (**), ψ must be the Gaussian composition of thė forms (a, b, c) and (a_1, b_1, c_1).

In the following we shall investigate a flow of primitive points on the sphere obtained with the help of rotations governed by one and the same class of forms (p), fixed by the prime number p. Theorem III.1.5 tells us that all such rotations are governed by the classes: $(p), (p)^2, (p)^3, \ldots,$ which becomes significant for further interpretations of the flow.

We shall make one more remark about rotations which will be useful for further generalizations to the domain of integral matrices. Let L, L' be two vectors with real components fulfilling the relations

$$L^2=L'^2=-m.$$

Then, as is easy to see,

$$(L+L')L=L'(L+L').$$

If the vector $B=L+L'$ is non-zero, that is,

$$L\neq -L',$$

then we have:

$$BLB^{-1}=L';$$

i.e., B governs the rotation (L, L').

§ 2. Asymptotic Properties of Rotations of a Large Sphere

We shall investigate rotations of the set of integral primitive points on the sphere (III.1.1) as the radius m increases to infinity. The number m

is chosen, as everywhere in this chapter, to satisfy one of the conditions $m \equiv 1, 2 \pmod 4$ or $m \equiv 3 \pmod 8$. Let us consider the integral primitive points on the sphere (III.1.1). For each such point we can find a primitive solution to the equation

(III.2.1) $$L^2 = -m.$$

The number $t(m)$ of such primitive points, as was elucidated in the introduction, satisfies the important inequalities of K. L. SIEGEL:

(III.2.2) $$c_1(\varepsilon)m^{\frac{1}{2}+\varepsilon} \geqq t(m) \geqq c_2(\varepsilon)m^{\frac{1}{2}-\varepsilon}$$

for every $\varepsilon > 0$.

In the following $\varepsilon, \varepsilon_1, \varepsilon_2, \ldots$ will be small positive constants, chosen successively, with each one depending on the previously chosen ones; $c_1, c_2, \ldots, c_1(\varepsilon_i), c_2(\varepsilon_i), \ldots$ are positive constants or positive constants depending on ε_i respectively.

Let r be a given odd integer such that

(III.2.3) $$(m, r) = 1, \quad (-m/p) = +1 \quad \text{for} \quad p \mid r.$$

Then for arbitrary integral $s \geqq 1$, the congruence

$$l^2 + m = 0 \pmod{r^s}$$

is solvable, and thus there exists a binary quadratic form (r^s, l, c) with determinant $(-m)$. If L is any primitive solution of (III.2.1) (in the sequel only such solutions will be considered), then $\mathrm{Norm}(l+L)$ will be divisible by r^s. Hence, according to the basic facts presented in Chapter II, we obtain

$$l + L = BV,$$

where $\mathrm{Norm}(B) = r^s$. If we write down such equations for all $t(m)$ of the primitive solutions of (III.2.1) $L_1, L_2, \ldots, L_{t(m)}$, then we get:

(III.2.4) $$\begin{cases} l + L_1 = B_1 V_1 \\ l + L_2 = B_2 V_2 \\ \quad \vdots \qquad \vdots \\ l + L_t = B_t V_t, \end{cases}$$

where $t = t(m)$ and the numbering of solutions is arbitrary. Corresponding to these relations we have the rotations $(L_i, B_i^{-1} L_i B_i)$, all of which are governed by the same form (r^s, l, c). If r^s is large, then the quaternions B_i have large norm. The study of such classes of rotations is fundamental in the sequel. Assuming r fixed, $m \to \infty$, we choose a fixed constant $\tau > 0$ as small as desired, and then we choose s such that

(III.2.5) $$\left(\frac{1}{2} + \tau\right) \frac{\ln m}{\ln r} \leqq s \leqq \left(\frac{1}{2} + \tau\right) \frac{\ln m}{\ln r} + 1$$

or, equivalently,

(III.2.6) $$m^{\frac{1}{2}+\tau} \leqq r^s \leqq r m^{\frac{1}{2}+\tau}.$$

The number l – a solution of the congruence $l^2+m\equiv 0 \pmod{r^s}$ – is selected to fulfill

$$\text{(III.2.7)} \qquad l^2+m\equiv 0 \pmod{r^s}, \quad \left(\frac{l^2+m}{r^s}, r\right)=1, \quad |l|\leqq r^{s+1}.$$

It is easy to show that this is possible.

Now an estimation theorem on the rotations of large spheres will be formulated, having basic character for the following.

Theorem III.2.1: *Let*

$$\text{(III.2.8)} \qquad \begin{aligned} l+L_1 &= B_1 V_1 \\ \vdots \quad & \quad \vdots \\ l+L_{t_1} &= B_{t_1} V_{t_1} \end{aligned}$$

be any $t_1 \leqq t$ *of the equations* (III.2.4), *where* s *and* l *are chosen as above, and*

$$t_1 > c_0 m^{\frac{1}{2}-\varepsilon_0},$$

where ε_0 *is a given small number. (Without loss of generality we may assume that these equations are the first* t_1 *equations of* (III.2.4).*) Then the number* w *of (not right associated) solutions* B_i *of these equations is bounded from below by the expression*

$$\text{(III.2.9)} \qquad w \geqq c_1(\varepsilon) m^{\frac{1}{2}-2\varepsilon_0-\varepsilon},$$

where $\varepsilon>0$ *is any positive number and* $c_1(\varepsilon)$ *is a constant depending only on* ε, r, c_0, ε_0.

The proof of this theorem, found also in LINNIK [10], requires the introduction of some new ideas and a series of approximations.

The pair (L_α, L_β) of integral primitive vectors of norm m will be called conjugate if it is possible to fulfil the equations

$$\text{(III.2.10)} \qquad \begin{aligned} l+L_\alpha &= B V_\alpha \\ l+L_\beta &= \bar{B} V_\beta, \end{aligned}$$

where l is the above-mentioned number and $\mathrm{Norm}(B)=r^s$. We shall prove that the number of conjugate pairs does not exceed

$$\text{(III.2.11)} \qquad c(\varepsilon) m^{\frac{1}{2}+\varepsilon}$$

for every $\varepsilon>0$. From this it is not hard to deduce the inequality (III.2.9). According to § 1 of this chapter, each conjugate pair (L_α, L_β) may be associated with a reduced positive binary form

$$\varphi=(a,b,c), \quad 2|b|\leqq c \leqq a, \quad b^2-ac=-m$$

governing the rotation (L_α, L_β). Moreover, in order to estimate the number of conjugate pairs, it is sufficient to find a bound for the number of reduced binary forms governing these rotations.

We divide the reduced positive binary forms (a, b, c) with determinant $(-m)$ into two sets:

a) the set of junior forms – forms for which the inequality

(III.2.12) $$c = \min \varphi > m^{\frac{1}{2} - (\tau - \frac{\tau^2}{4})}$$

holds.

b) the set of senior forms, for which we have

(III.2.13) $$c = \min \varphi \leqq m^{\frac{1}{2} - (\tau - \frac{\tau^2}{4})}.$$

Let us first look at the conjugate pairs governed by junior forms. We shall show that the number of conjugate pairs governed by junior forms does not exceed the quantity given in (III.2.11).

1. First of all we prove that for sufficiently large m and any L_α there exists at most one L_β such that (L_α, L_β) forms a conjugate pair governed by a junior form. Indeed, let (L_α, L_β) and (L_α, L_γ) be two conjugate pairs with

(III.2.14) $$\begin{aligned} l + L_\alpha &= B V_\alpha \\ l + L_\beta &= \bar{B} V_\beta \\ l + L_\gamma &= \bar{B} V_\gamma , \end{aligned}$$

and let the pairs (L_α, L_β) and (L_α, L_γ) be governed by the junior forms (a_1, b_1, c_1) and (a_2, b_2, c_2) respectively. Then the following inequalities are valid:

(III.2.15) $$\begin{aligned} 2|b_1| \leqq c_1 \leqq a_1, \ m^{\frac{1}{2} - (\tau - \frac{\tau^2}{4})} < c_1 \leqq (4m/3)^{\frac{1}{2}} \\ 2|b_2| \leqq c_2 \leqq a_2, \ m^{\frac{1}{2} - (\tau - \frac{\tau^2}{4})} < c_2 \leqq (4m/3)^{\frac{1}{2}} . \end{aligned}$$

Since (a_1, b_1, c_1) and (a_2, b_2, c_2) govern the rotations (L_α, L_β) and (L_α, L_γ) respectively, integral quaternions A_1, C_1, A_2, C_2 can be found such that

(III.2.16) $$\begin{aligned} l_1 + L_\alpha = A_1 C_1, \ \mathrm{Norm}(A_1) = a_1, \ \mathrm{Norm}(C_1) = c_1 , \\ C_1 L_\alpha C_1^{-1} = \bar{A}_1 L_\alpha \bar{A}_1^{-1} = L_\beta , \\ l_2 + L_\alpha = A_2 C_2, \ \mathrm{Norm}(A_2) = a_2, \ \mathrm{Norm}(C_2) = c_2 , \\ C_2 L_\alpha C_2^{-1} = \bar{A}_2 L_\alpha \bar{A}_2^{-1} = L_\gamma . \end{aligned}$$

a) We can show now that

(III.2.17) $$\begin{aligned} 2\mathscr{R}(BC_1) \equiv 2\mathscr{R}(BC_2) \equiv 2\mathscr{R}(B\bar{A}_1) \equiv \\ \equiv 2\mathscr{R}(B\bar{A}_2) \equiv 0 \ (\mathrm{mod}\, r^s) , \end{aligned}$$

using (III.2.14)—(III.2.16). Indeed, we have

(III.2.18) $$C_1(l + L_\alpha)C_1^{-1} = l + L_\beta, \quad C_1 B V_\alpha = \bar{B} V_\beta C_1 .$$

Because $(\mathrm{Norm}(V_\alpha), \mathrm{Norm}(\bar{B})) = 1$, there exist integral quaternions X, Y with

$$\bar{B}X + V_\alpha Y = 1 .$$

From (III.2.18) it follows that

$$\begin{aligned} C_1 B = C_1 B \cdot 1 = C_1 B(\bar{B}X + V_2 Y) = r^s C_1 X + C_1 B V_2 Y \\ = \bar{B} B C_1 X + \bar{B} V_\beta C_1 Y = \bar{B} C_1' . \end{aligned}$$

Likewise, the corresponding equations for C_2, A_1, A_2 follow and we have:

(III.2.19) $C_1 B = \bar{B} C_1', \quad C_2 B = \bar{B} C_2', \quad \bar{A}_1 B = \bar{B} A_1', \quad \bar{A}_2 B = \bar{B} A_2'$.

Hence, $C_1, C_2, \bar{A}_1, \bar{A}_2$ belong to the left versored ray (mod $\bar{B}$), which implies (III.2.17).

b) We shall now prove that for sufficiently large m the quaternions BC_1 and BC_2 are vectors. We have

$$|2\mathscr{R}(BC_1)| \leqq 2(\operatorname{Norm}(B)\operatorname{Norm}(C))^{\frac{1}{2}} < \\ < 2(rm^{\frac{1}{2}+\tau}(4m/3)^{\frac{1}{2}})^{\frac{1}{2}} < c_1 m^{\frac{1}{2}+\frac{\tau}{2}}.$$

In view of the congruences (III.2.17), $2\mathscr{R}(BC_1)$ is divisible by r^s, and

$$r^s > m^{\frac{1}{2}+\tau}.$$

Hence it follows for sufficiently large m that $2\mathscr{R}(BC_1)=0$; i.e., BC_1 is a vector. Analogously one proves that BC_2 is a vector.

We note that in **a)** and **b)** above we have not made use of the fact that the forms (a_1, b_1, c_1) and (a_2, b_2, c_2) are junior forms, so that these results remain true in the case of senior forms also.

c) We shall show that $B\bar{A}_1$ and $B\bar{A}_2$ are also vectors for sufficiently large m. Here it will be necessary to use the relations (III.2.15); i.e., the fact that we are dealing with junior forms is brought into play. We have:

$$a_1 c_1 - b_1^2 = m, \quad |a_1| = \frac{|b_1^2| + m}{|c_1|} < \\ < \frac{(\frac{1}{2}(\frac{4}{3}m)^{\frac{1}{2}})^{\frac{1}{2}} + m}{m^{\frac{1}{2}-(\tau-\frac{\tau^2}{4})}} < 2m^{\frac{1}{2}+\tau-\frac{\tau^2}{4}}.$$

Hence

$$|2\mathscr{R}(B\bar{A}_1)| \leqq 2\sqrt{\operatorname{Norm}(B)\operatorname{Norm}(A_1)} \leqq \\ \leqq 2(rm^{\frac{1}{2}+\tau} 2m^{\frac{1}{2}+\tau-\frac{\tau^2}{4}})^{\frac{1}{2}} < cm^{\frac{1}{2}+\tau-\frac{\tau^2}{8}},$$

so that for m large enough, $2\mathscr{R}(B\bar{A}_1)=0$ and $B\bar{A}_1$ is a vector. A similar argument shows that $B\bar{A}_2$ is a vector.

d) For sufficiently large m we can regard the quaternions BC_1, BC_2, $B\bar{A}_1$, $B\bar{A}_2$ as vectors with three components having the orthogonal directions i, j, k. Let us consider the products

$$(B\bar{A}_1)(BC_1) = (A_1\bar{B})(BC_1) = -r^s A_1 C_1 = -b_1 r^s - r^s L_\alpha, \\ (B\bar{A}_2)(BC_2) = -(A_2\bar{B})(BC_2) = -r^s A_2 C_2 = -b_2 r^s - r^s L_\alpha.$$

It follows from these equations that the vectors $B\bar{A}_1$, BC_1, $B\bar{A}_2$, BC_2 are perpendicular to the vector L_α and consequently, they lie in the same plane. Then we can find real numbers λ and μ such that

$$BC_2 = \lambda B\bar{A}_1 + \mu BC_1,$$

or

$$C_2 = \lambda \bar{A}_1 + \mu C_1.$$

But we also have the equations

$$\bar{A}_1 L_\alpha = L_\beta \bar{A}_1, \quad C_1 L_\alpha = L_\beta C_1\,;$$

hence we conclude:

$$C_2 L_\alpha = (\lambda \bar{A}_1 + \mu C_1) L_\alpha = L_\beta (\lambda \bar{A}_1 + \mu C_1) = L_\beta C_2\,,$$

that is, $C_2 L_\alpha C_2^{-1} = L_\beta$ and $L_\beta = L_\gamma$, which is the assertion we were trying to prove.

e) In view of the Siegel inequalities (III.2.2), the number $t(m)$ of vectors L_α does not exceed

$$c(\varepsilon) m^{\frac{1}{2}+\varepsilon}.$$

To each vector L_α corresponds, because of point **1** above, no more than one conjugate pair (L_α, L_β) governed by a junior form. This means that the number of such pairs does not exceed (III.2.11), which was the first step in our proof.

More complicated is the approximation to the number of conjugate pairs governed by senior forms. We shall prove that this number does not exceed

$$c(\varepsilon) m^{\frac{1}{2} - \tau + \frac{\tau^2}{8} + \varepsilon}. \tag{III.2.20}$$

1) We introduce the concept of index of imprimitivity for conjugate pairs. Let (L_α, L_β) be a conjugate pair governed by the senior form (a,b,c), so that the following equations are valid:

$$\begin{cases} l + L_\alpha = B V_\alpha, \quad l + L_\beta = \bar{B} V_\beta \\ C L_\alpha C^{-1} = L_\beta,\ l + L_\alpha = AC,\ \mathrm{Norm}(A) = a,\ \mathrm{Norm}(C) = c\,, \\ 2|b| \leqq c \leqq a,\ c \leqq m^{\frac{1}{2} - (\tau - \frac{\tau^2}{4})}. \end{cases} \tag{III.2.21}$$

It is possible that the quaternion CB is not primitive. Because C and B are primitive, by Theorem III.1.2 we can find a quaternion T such that

$$\begin{cases} C = C'T,\ B = \bar{T} B',\ C'B' \ \ \text{primitive}\,, \\ \mathrm{Norm}(T) = r^{s_1}, \quad \text{where} \quad 0 \leqq s_1 \leqq s\,. \end{cases} \tag{III.2.22}$$

The quantity r^{s_1} will be called the index of imprimitivity of the pair (L_α, L_β).

First we shall prove the following statement: the number of conjugate pairs (L_α, L_β) with given index of imprimitivity r^{s_1} governed by the given form (a, b, c) with determinant $(-m)$ satisfying the conditions

$$2|b| \leqq c \leqq a, \quad 1/2 \cdot m^{\frac{1}{2} - \nu} < c \leqq m^{\frac{1}{2} - \nu}, \tag{III.2.23}$$

where ν is any number for which

$$\nu \geqq \tau - \tau^2/4\,, \tag{III.2.24}$$

does not exceed

$$c(\varepsilon) m^{\frac{\nu - \tau}{2} + \varepsilon} r^{s_1} e^{\frac{1}{2}}, \tag{III.2.25}$$

where $e = (c, m)$.

The proof of this statement is rather long.

a) Let the conjugate pair (L_α, L_β) with index of imprimitivity r^{s_1} be given so that (III.2.21) and (III.2.22) are valid. We examine the quaternion

$$L'_\alpha = T L_\alpha T^{-1},$$

finding

$$L'^2_\alpha = T L^2_\alpha T^{-1} = T(-m)T^{-1} = -m,$$

which means that L'_α is a vector with norm m. Further,

$$L'_\alpha = \bar{T}^{-1} L_\alpha \bar{T},\ l + L'_\alpha = l + \bar{T}^{-1} L_\alpha \bar{T} = B' V_\alpha \bar{T} = B' V'_\alpha,$$

which shows that L'_α is an integral vector; we have

$$L_\alpha = T^{-1} L'_\alpha T = \frac{\bar{T} L'_\alpha T}{\mathrm{Norm}(T)} = \frac{\bar{T} L'_\alpha T}{r^{s_1}}.$$

If L'_α were not primitive, then its divisor would also divide r^{s_1}, and this is not possible because of $(m, r) = 1$.

Thus, L'_α is an integral primitive vector with norm m. The following relations are valid:

$$\text{(III.2.26)}\qquad \begin{aligned} &L'_\alpha = T L_\alpha T^{-1} = \bar{T}^{-1} L_\alpha \bar{T},\ L_\alpha = T^{-1} L'_\alpha T,\\ &l + L'_\alpha = B' V'_\alpha, \quad \text{where} \quad V'_\alpha = V_\alpha T,\\ &C' L'_\alpha C'^{-1} = L_\beta\\ &l + L'_\alpha = A' C', \quad \text{where} \quad \mathrm{Norm}(A') = a',\ \mathrm{Norm}(C') = c'. \end{aligned}$$

b) Points **a)** and **b)** of the preceding part did not require the form under consideration to be junior. Following those considerations it can be shown that for sufficiently large m,

$$2\mathscr{R}(C' B') = 2\mathscr{R}(B' C') = \frac{2}{r^{s_1}} \cdot \mathscr{R}(B C) = 0;$$

accordingly, $C' B'$ is a vector and we have

$$\text{(III.2.27)}\qquad C' B' = -B' C'.$$

c) Now we shall prove

$$\text{(III.2.28)}\qquad 2\mathscr{R}(V'_\alpha \bar{C'}) \equiv 0 \pmod{c'},$$

where $c' = \mathrm{Norm}(C')$. In fact, taking into account the equations (III.2.26) and (III.2.27), we obtain

$$\begin{aligned} \bar{B'} \bar{C'} V'_\alpha \bar{C'} &= -C'(B' V'_\alpha) \bar{C'} = -C'(l + L'_\alpha) \bar{C'}\\ &= -l\,\mathrm{Norm}(C') - \mathrm{Norm}(C') \cdot C' L'_\alpha C'^{-1} = (\mathrm{Norm}(C'))(-l - L_\beta), \end{aligned}$$

so that

$$\text{(III.2.29)}\qquad \bar{B'} \bar{C'} V'_\alpha \bar{C'} \equiv 0 \pmod{c'}.$$

Since *right g.c.d.* $(\bar{B'}, C') = 1$ (the quaternion $C' B'$ is primitive), there exist integral quaternions X and Y such that $X\bar{B'} + Y C' = 1$. Using this equation, we can write

$$\begin{aligned} \bar{C'} V'_\alpha \bar{C'} &= (X \bar{B'} + Y C')(\bar{C'} V'_\alpha \bar{C'})\\ &= X \bar{B'} \bar{C'} V'_\alpha \bar{C'} + Y C' \bar{C'} V'_\alpha \bar{C'} = c' V''_\alpha. \end{aligned}$$

Therefore, $V'_\alpha \overline{C'} = C' V''_\alpha$, and V'_α belongs thus to the left versored ray $(\operatorname{mod} C')$; the congruence (III.2.28) is valid.

d) We set:

$$\text{(III.2.30)}\quad \begin{aligned} &C'B' = g'_1 i + g'_2 j + g'_3 k = H_1, \quad H_1 \text{ primitive},\\ &V'_\alpha \overline{C'} = dc'/2 + g''_1 i + g''_2 j + g''_3 k = H_2, \quad \text{where } d \text{ is integral},\\ &\operatorname{Norm}(H_1) = \operatorname{Norm}(C')\operatorname{Norm}(B') = c' r^{s-s_1} = h_1,\\ &\operatorname{Norm}(H_2) = \operatorname{Norm}(V'_\alpha)\operatorname{Norm}(\overline{C'}) = v'c' = h_2 . \end{aligned}$$

Consider now the form

$$\operatorname{Norm}(-\overline{H}_1 x + H_2 y).$$

On the one hand, applying the formula $\operatorname{Norm}(A+B) = \operatorname{Norm}(A) + \operatorname{Norm}(B) + 2\mathscr{R}(\overline{A}B)$, we get

$$\begin{aligned} \operatorname{Norm}(-\overline{H}_1 x + H_2 y) &= \operatorname{Norm}(-\overline{H}_1 x) + 2\mathscr{R}(-H_1 x H_2 y) + \operatorname{Norm}(H_2 y)\\ &= h_1 x^2 - 2xyc'\mathscr{R}(V'_\alpha B') + h_2 y^2\\ &= h_1 x^2 - 2c'lxy + h_2 y^2 , \end{aligned}$$

and in view of (III.2.26),

$$\mathscr{R}(V'_\alpha B') = \mathscr{R}(B' V'_\alpha) = l .$$

On the other hand,

$$\begin{aligned} \operatorname{Norm}(-\overline{H}_1 x + H_2 y) = \left(\frac{dc'}{2}\right)^2 y^2 + (g'_1 x + g''_1 y)^2 +\\ + (g'_2 x + g''_2 y)^2 + (g'_3 x + g''_3 y)^2 . \end{aligned}$$

Equating these expressions, we obtain

$$\text{(III.2.31)}\quad \begin{aligned} &4h_1 x^2 - 2(4c'l)xy + (4h_2 - d^2 c'^2) y^2\\ &= (2g'_1 x + 2g''_1 y)^2 + (2g'_2 x + 2g''_2 y)^2 + (2g'_3 x + 2g''_3 y)^2 . \end{aligned}$$

Equation (III.2.31) is a representation of the binary form $(4h_1, -4c'l, 4h_2 - d^2c'^2)$ as a sum of three squares. This representation possesses the following two properties:

$$(2g'_1, 2g'_2, 2g'_3) = 2,$$

$$(4h_1, -4c'l, 4h_2 - d^2c'^2) \,\Big|\, \left(\begin{vmatrix} 2g'_2 & 2g'_3\\ 2g''_2 & 2g''_3 \end{vmatrix}, \begin{vmatrix} 2g'_3 & 2g'_1\\ 2g''_3 & 2g''_1 \end{vmatrix}, \begin{vmatrix} 2g'_1 & 2g'_2\\ 2g''_1 & 2g''_2 \end{vmatrix} \right).$$

The first of these properties is obvious, since H_1 is a primitive vector. We shall prove the second one. It is true that

$$(4h_1, -4c'l, 4h_2 - d^2c'^2) = c'(4r^{s-s_1}, -4l, 4v' - d^2c'^2) = 4c',$$

since $(r, l) = 1$. Further,

$$\begin{aligned} 4c'V'B' &= (2V'_\alpha \overline{C'})(2C'B) = 2H_1 \cdot 2H_2\\ &= \mathscr{R}(H_1 H_2) + (2dc'g'_1 + 2g'_2\, 2g''_3 - 2g'_3\, 2g''_2) i +\\ &\quad + (2dc'g'_2 + 2g'_3\, 2g''_1 - 2g'_1\, 2g''_3) j +\\ &\quad + (2dc'g'_3 + 2g'_1\, 2g''_2 - 2g'_2\, 2g''_1) k . \end{aligned}$$

Hence the second property follows immediately.

e) We observe next that to given $d, g'_1, g'_2, g'_3, g''_1, g''_2, g''_3$ there correspond no more than 24^2 conjugate pairs (L_α, L_β). The above quantities are determined by the quaternions H_1 and H_2. From the equation $H_1 = C'B'$ we see then that the quaternion C' is uniquely determined up to a unit factor. Hence we have no more than 24 distinct pairs C', B'. But for given C', V'_α is uniquely determined by the equation $H_2 = V'_\alpha \overline{C'}$, L'_α by the equation $l + L'_\alpha = B' V'_2$, and L_β by the equation $C' L'_\alpha C'^{-1} = L_\beta$. Finally, because of $l + L_\alpha = B V_\alpha = \overline{T} B' V_\alpha$, $l + L'_\alpha = B' V T$, the quaternion T is determined up to a unit factor. Therefore, to each C' corresponds exactly one L_β and not more than 24 vectors L_α; this proves the above statement.

f) Thus, in order to estimate the number of conjugate pairs (L_α, L_β) governed by the given senior form (a, b, c) and having the given index of imprimitivity r^{s_1}, it is sufficient to estimate the number of distinct pairs H_1 and H_2.

Given (a, b, c) and r^{s_1}, the numbers $h_1 = c' r^{s-s_1} = c r^{s-2s_1}$ and $h_2 = v'c' = vc = \dfrac{l^2+m}{r^s} c$ are completely determined. The number of distinct d is estimated by the inequality

$$|d| < 2\sqrt{2}\, m^{\frac{1}{2}-\frac{1}{2}} r^{s_1}. \tag{III.2.32}$$

In view of the equation (III.2.31), the form $(4h_1, -4lc', 4h_2 - d^2 c'^2)$ is non-negative, and thus

$$4h_1(4h_2 - d^2 c'^2) - (4lc')^2 \geqq 0.$$

Substituting for h_1 and h_2 the relations $h_1 = c' r^{s-s_1}$, $h_2 = v'c'$, we get

$$4(l^2 - r^{s-s_1} v') + r^{s-s_1} d^2 c' \leqq 0.$$

But $l + L'_\alpha = B' V'_\alpha$ and $l^2 - r^{s-s_1} v' = -m$; therefore, using (III.2.23),

$$|d| \leqq \left(\frac{4m}{r^{s-s_1} c'}\right)^{\frac{1}{2}} = \left(\frac{4m}{r^{s-2s_1} c}\right)^{\frac{1}{2}} < 2\sqrt{2}\, m^{\frac{v-\tau}{2}} r^{s_1},$$

which coincides with the inequality (III.2.32).

Further, for given d, the number of distinct systems $g'_1, g'_2, g'_3, g''_1, g''_2, g''_3$, in virtue of (III.2.31) (under the two conditions following that equation) and Theorem III.1.3 (the estimate (III.1.8)), does not exceed

$$\begin{aligned} & c(\varepsilon) D^\varepsilon \left[\left(c', \frac{D}{c'^2}\right)\right]^{\frac{1}{2}} \\ & = c(\varepsilon) \left[4c'^2 \left(4md^2 c' \frac{r^s}{r^{s_1}}\right)\right]^\varepsilon \left[\left(c', 16m - 4d^2 c' \frac{r^s}{r^{s_1}}\right)\right]^{\frac{1}{2}} < \\ & < c'(\varepsilon) m^\varepsilon [(c', m)]^{\frac{1}{2}} = c'(\varepsilon) m^\varepsilon e^{\frac{1}{2}}, \end{aligned} \tag{III.2.33}$$

where D stands for the determinant of the form

$$(4h_1, -4c'l, 4h_2 - d^2c'^2); \quad e = (c', m) = (c, m).$$

This justifies, at last, the estimate (III.2.25).

2) Now we shall derive (III.2.20) from (III.2.25). Here it will be necessary for us to compute the number of distinct senior forms. For this we divide the segment from 0 to $m^{\frac{1}{2}-\tau+\frac{\tau^2}{4}}$ into $n < c \ln m < c(\varepsilon)m^{\varepsilon}$ intervals by the points

$$m^{\frac{1}{2}-\nu_1} = \tfrac{1}{2} m^{\frac{1}{2}-\tau+\frac{\tau^2}{4}}, \quad m^{\frac{1}{2}-\nu_2} = \tfrac{1}{2} m^{\frac{1}{2}-\nu_1}, \tag{III.2.34}$$
$$m^{\frac{1}{2}-\nu_3} = \tfrac{1}{2} m^{\frac{1}{2}-\nu_2}, \ldots, m^{\frac{1}{2}-\nu_n} = \tfrac{1}{2} m^{\frac{1}{2}-\nu_{n-1}}$$

so that if $m^{\frac{1}{2}-\nu_n} \geqq \frac{1}{2}$, then $\frac{1}{2} m^{\frac{1}{2}-\nu_n} \leqq \frac{1}{2}$.

The number of conjugate pairs (L_α, L_β) with given index of imprimitivity r^{s_1} which are governed by some senior form (a,b,c) under the condition

$$\tfrac{1}{2} m^{\frac{1}{2}-\nu_\gamma} < c \leqq m^{\frac{1}{2}-\nu_\gamma}$$

is not larger than the number of pairs (L_α, L_β) with index of imprimitivity r^{s_1} governed by a given form (a, b, c) under the above-mentioned condition on c, multiplied by the number of different forms (a, b, c) which satisfy the condition and govern conjugate pairs with indices of imprimitivity r^{s_1}. The first number is bounded from above by the expression in (III.2.25). The second quantity may be estimated with the help of Theorem III.1.4 (expression (III.1.10)). From this theorem it easily follows that the second quantity does not exceed

$$c(\varepsilon)\frac{m^{\frac{1}{2}-\nu_\gamma+\varepsilon}}{r^{s_1} e^{\frac{1}{2}}}. \tag{III.2.35}$$

Uniting (III.2.25) and (III.2.35), we see that the number of conjugate pairs (L_α, L_β) with index of imprimitivity r^{s_1} governed by an senior form under the above-mentioned condition does not exceed

$$\left(c'(\varepsilon) m^{\frac{1}{2}\nu_\gamma - \frac{1}{2}\tau+\varepsilon} r^{s_1} e^{\frac{1}{2}}\right)\left(c(\varepsilon) m^{\frac{1}{2}-\nu_\gamma+\varepsilon} r^{-s_1} e^{-\frac{1}{2}}\right) \leqq \tag{III.2.36}$$
$$\leqq c(\varepsilon) m^{\frac{1}{2}-\tau+\frac{\tau^2}{8}+\varepsilon},$$

since $\nu_\gamma \geqq \tau - \tau^2/4$.

Finally, the number of different ν_γ and the number of distinct indices of imprimitivity do not exceed $c(\varepsilon)m^{\varepsilon}$. Bringing this together with (III.2.36) produces at last the approximation in (III.2.20). For junior forms we have obtained the estimate $c(\varepsilon)m^{\frac{1}{2}+\varepsilon}$. Putting this together with (III.2.36) we get that the total number of conjugate pairs does not exceed

$$c(\varepsilon) m^{\frac{1}{2}+\varepsilon}. \tag{III.2.37}$$

Now it is easy to prove Theorem III.2.1 (the lower approximation (III.2.9)). Returning to the equations (III.2.8), we write them in the form (changing the order if necessary):

$$\begin{aligned} l+L_\alpha &= B_1 V_\alpha \quad (\alpha = 1, \ldots, n_1) \\ l+L_\alpha &= B_2 V_\alpha \quad (\alpha = n_1+1, \ldots, n_1+n_2) \\ &\vdots \\ l+L_\alpha &= B_w V_\alpha \quad (\alpha = n_1+\cdots+n_{w-1}+1, \ldots, n_1+\cdots+n_w), \end{aligned}$$

where $B_1, B_2, \ldots, B_w$ are all different. By hypothesis,

(III.2.38) $$t_1 = n_1 + \cdots + n_w > c_0 m^{\frac{1}{2}-\varepsilon_0}.$$

Now we construct $n_1^2 + \cdots + n_w^2$ different conjugate pairs (L_α, L'_β) in the following manner. Let

$$l+L_\alpha = B_\gamma V_\alpha, \quad l+L_\beta = B_\gamma V_\beta,$$

where for given $\gamma = 1, 2, \ldots, w$, the numbers α and β are chosen from $n_1+\cdots+n_{\gamma-1}+1, \ldots, n_1+\cdots+n_{\gamma-1}+n_\gamma$ (α and β may be equal). Passing in the second equation to conjugates, we obtain

$$l-L_\beta = \bar{V}_\beta \bar{B}_\gamma, \quad l+\bar{L}'_\beta = \bar{B}_\gamma V'_\beta,$$

where

$$L'_\beta = -\bar{B}_\gamma L_\beta \bar{B}_\gamma^{-1}, \quad V'_\beta = \bar{V}_\beta.$$

Here, L'_β is an integral primitive vector with norm m, and obviously, different L_β will produce different L'_β. Therefore, we get $n_1^2 + \cdots + n_w^2$ distinct conjugate pairs (L_α, L'_β). But, as we proved, the number of such pairs cannot exceed $c(\varepsilon) m^{\frac{1}{2}+\varepsilon}$ (see (III.2.37)). Therefore, making use of the inequality

$$w(n_1^2 + \cdots + n_w^2) \geqq (n_1 + \cdots + n_w)^2,$$

we get

$$c(\varepsilon) m^{\frac{1}{2}+\varepsilon} \geqq n_1^2 + \cdots + n_w^2 \geqq \frac{1}{w}(n_1 + \cdots + n_w)^2 = \frac{t_1^2}{w} > \frac{c^2 m^{1-2\varepsilon_0}}{w},$$

in view of (III.2.38). It follows that

$$w \geqq \frac{c_0^2}{c(\varepsilon)} m^{\frac{1}{2}-2\varepsilon_0-\varepsilon}$$

and Theorem III.2.1 is proved.

Far-reaching generalizations of Theorem III.2.1, dealing with transformations of large ellipsoids of sufficiently arbitrary form into themselves and into other ellipsoids, instead of rotations of large spheres, have been obtained by A. V. MALYSHEV. These generalizations provide the possibility of constructing a complete theory of asymptotic solutions of the diophantine equations

(III.2.39) $$f(x, y, z) = m,$$

where $f(x, y, z)$ is a positive ternary quadratic form.

Chapter IV

Asymptotic Geometrical and Ergodic Properties of the Set of Integral Points on the Sphere

§ 1. Formulation of the Problem

In this chapter we shall apply the results of the preceding chapters on rotations of the sphere and their governing binary quadratic forms to problems concerning the asymptotic distribution of integral points on the sphere as the radius of the sphere increases to infinity.

We shall consider integral points of $\mathrm{Sp}_3(m)$: $x^2+y^2+z^2=m$, where m is an integer and $m \equiv 1, 2 \pmod 4$ or $m \equiv 3 \pmod 8$. We shall project the sphere $\mathrm{Sp}_3(m)$ centrally onto $\mathrm{Sp}_3(1)$: $x^2+y^2+z^2=1$, and study the asymptotic distribution of integral primitive points (x, y, z) on $\mathrm{Sp}_3(m)$. The problem concerning the distribution of integral points is easily reduced to results in the preceding chapters.

Let Γ_0 be a closed convex region on $\mathrm{Sp}_3(1)$ with a piecewise smooth boundary, Γ its central projection in $\mathrm{Sp}_3(m)$, $H(\Gamma)$ the number of integral points on $\mathrm{Sp}_3(m)$ contained in Γ, $H_0(\Gamma)$ the number of primitive points of this type.

If $\Gamma_0=\mathrm{Sp}_3(1)$, $\Gamma=\mathrm{Sp}_3(m)$, then $H_0(\Gamma)$ represents the total number of primitive points on $\mathrm{Sp}_3(m)$, which we denote then by $H_0(m)$; correspondingly, $H(\mathrm{Sp}_3(m))$ is denoted by $H(m)$. According to the theorem of K. L. Siegel (inequality (III.2.2)),

$$\ln H_0(m) \sim \tfrac{1}{2} \ln m \qquad \text{(IV.1.1)}$$

as $m \to \infty$ and $m \equiv 1, 2, 3, 5, 6 \pmod 8$ (only such values of m will be considered in the following). This result of K. L. Siegel, characterizing the growth of the number of integral points on the sphere with increasing radius, is proved by a deep and difficult method. The theorems given here are complements to this result of geometric and ergodic types.

Theorem IV.1.1: *Let $q \geqq 3$ be a prime such that $(-m/q)=+1$ and let Γ be a region of $\mathrm{Sp}_3(m)$ as above. Denote by $\omega(\Gamma)$ the solid angle covered by the surface Γ, measured from the center of the sphere. Then for $m \to \infty$ we have:*

$$H_0(\Gamma)=\frac{\omega(\Gamma)}{4\pi} H_0(m)\,(1+\kappa_0(\Gamma_0, q, m)), \qquad \text{(IV.1.2)}$$

$$H(\Gamma)=\frac{\omega(\Gamma)}{4\pi} H(m)\,(1+\kappa(\Gamma_0, q, m)), \qquad \text{(IV.1.3)}$$

where $\kappa_0(\Gamma_0, q, m)$ and $\kappa(\Gamma_0, q, m)$ approach zero for fixed Γ_0, q as $m \to \infty$.

In this theorem the auxiliary number q appears; its presence is apparently explained by the deficiency in method and not essential (we encounter another situation in "ergodic" theorems). However, if we do not introduce the extraneous number q, then it is only possible to prove some conditional theorems (see Chapter IX).

§ 2. Ergodic Properties

We shall now turn to ergodic properties of the set of primitive integral points on $\mathrm{Sp}_3(m)$. If $L=(x, y, z)$ is one such point, then, by reflecting it in the coordinate planes and by revolving it through 120° around the coordinate "bisectors", we obtain new primitive points.

Let us look at the spherical triangle in $\mathrm{Sp}_3(m)$ bounded by the intersection of $\mathrm{Sp}_3(m)$ with the planes $z=0$, $y-z=0$, $x-z=0$, in which the first two sides are included, the third side rejected. We shall call this fundamental region Ω.

Let $q \geqq 3$ be prime with $(-m/q)=+1$, and denote by k an arbitrary natural number. We consider rational revolutions and reflections of three-dimensional space of order q^k. By this we mean orthogonal matrices T of third order with $\det(T)=+1$, having the form $T=(\alpha_{ij})$, where $\alpha_{ij}=a_{ij}/q^k$, a_{ij} integral, and where at least one of the fractions α_{ij} is irreducible.

As a result of the revolution T an integral primitive point $L=(x, y, z) \in \mathrm{Sp}_3(m)$ will be carried into a point which we denote by $L'=TL$. This point is generally not integral. In the following we prove a simple lemma about revolutions T carrying $L \in \Omega$ into the integral point $L'=TL \neq L$ lying in Ω. As it turns out, there exist for each L exactly two such revolutions, depending on the choice of a solution to the congruence $\zeta_0^2+m \equiv 0 \pmod{q^k}$ (if $(-m/q)=+1$ there are exactly two solutions ζ_0 and $(-\zeta_0) \pmod{q^k}$ to this congruence). Here it is essential that L' and L lie in Ω. Also, L' is primitive if L is primitive. Choosing a definite representation for the solutions, one for each primitive point $L \in \mathrm{Sp}_3(m)$, T is then uniquely determined along with the image $L'=TL$ of L, which is also primitive. Thus, $TL=L'$ carries the primitive point L into the primitive point L'. We shall say that the operation T defines a flow on the set A of primitive points of $\Omega \subset \mathrm{Sp}_3(m)$. We denote the repetition of T r times by T^r. Project the spherical triangle Ω into $\mathrm{Sp}_3(1)$ and denote by Ω_0 the projection. The set A will then be transformed into a set A_0 and the flow goes over into a flow on A_0. We shall be concerned with this flow.

Let $\Lambda_0 \subset \Omega_0$, Ω_0 being the region defined above, Λ_0 having a piecewise smooth boundary. Denote by $f_{\Lambda_0}(X)$ the indicator function of the set Λ_0: $f_{\Lambda_0}(X)=1$ if $X \in \Lambda_0$ and $f_{\Lambda_0}(X)=0$ if $X \notin \Lambda_0$. The arguments of this function will be projections of primitive points.

The points $X, TX, T^2X, \ldots, T^rX$ form a trajectory of our flow. We shall be concerned with the ergodic mean of $f_{\Lambda_0}(X)$ (see Chapter I).

Let $\omega(\Lambda_0)$ be the solid angle spanned by the set Λ_0.

Theorem IV.2.1 *(ergodic theorem):*

$$\text{(IV.2.1)}\quad \frac{f_{\Lambda_0}(X)+f_{\Lambda_0}(TX)+\cdots+f_{\Lambda_0}(T^{s-1}X)}{s}=\frac{6\omega(\Lambda_0)}{\pi}(1+\kappa(q^k,\Lambda_0,m))$$

for all images X of primitive points, with the possible exception of $o(H(m))$ *such points;* $s \geqq c_0 \ln m$, *where* $\kappa(q^k, \Lambda_0, m) \to 0$ *for fixed* q^k, Λ_0, *as* $m \to \infty$.

Let M_0 be any set of points projected from A to the unit sphere $\mathrm{Sp}_3(1)$. We shall denote by T^rM_0 the set of points obtained after the r-fold primitive transformation T^r; by $\mathcal{M}(M_0)$ the number of points of M_0; $\mathcal{M}(T^lM_0 \cap \Lambda_0)$ is the number of points in $T^lM_0 \cap \Lambda_0$.

Theorem IV.2.2 *(mixing theorem): Let* $l=0, 1, 2, \ldots, s \geqq c_1 \ln m$, *and* $\mathcal{M}(M_0) > \varepsilon_0 H_0(m)$ $(\varepsilon_0 > 0$ *fixed). Then for every index l, with the possible exception of* $s \cdot o(1)$ $(m \to \infty)$ *of these indices, we have*

$$\text{(IV.2.2)}\qquad \mathcal{M}(T^lM_0 \cap \Lambda_0) = 6/\pi\ \omega(\Lambda_0)\mathcal{M}(M_0)(1+\kappa(q^k, \Lambda_0, \varepsilon_0, m)),$$

where $\kappa(q^k, \Lambda_0, \varepsilon_0, m) \to 0$ *for given* q^k, Λ_0, ε_0 *as* $m \to \infty$.

We remark that Theorem IV.1.1 on the asymptotic uniform distribution of integral points on the sphere turns out to be a special case of the mixing Theorem IV.2.2.

Turning to the proof of Theorem IV.1.1, we shall make use of results given in Chapters II and III and also of the terminology of those chapters. Once more we explain shortly most of the properties we shall use.

Let L and L' be primitive vectors of norm m. Then there exist integral quaternions A, C and an integer b such that

$$\text{(IV.2.3)}\qquad b+L=AC,\quad CLC^{-1}=\bar{A}L\bar{A}^{-1}=L',\quad b+L'=CA.$$

The ordered pair (L, L') will be called a rotation. The equations (IV.2.3) determine the rotation; however, to each such rotation correspond many sets of equations. We associate with the rotation (L, L') the following positive binary quadratic form of determinant $(-m)$:

$$(a, b, c) = ax^2+2bxy+cy^2 = \mathrm{Norm}(\bar{A}x+Cy),$$

so that

$$a=\mathrm{Norm}(A),\quad c=\mathrm{Norm}(C),\quad b=\mathcal{R}(AC).$$

We shall say that this form governs the rotation (L, L'). There may be many such forms. The set of these forms for given L and L' is a class of forms of determinant $(-m)$. If $m \equiv 1, 2 \pmod 4$, then this class is properly primitive; if $m \equiv 3 \pmod 8$, then one of the classes belonging to (L, L') and $(L, (1+i)^{-1}L'(1+i))$ is properly primitive and the other is improperly primitive. Further, let L be a given primitive vector of norm $m>3$ and, for $m \equiv 1, 2, \pmod 4$ let a class of properly primitive forms of determinant $(-m)$ be given, or for $m\equiv 3 \pmod 8$ a class of improperly primitive forms with the same determinant. Then there exist exactly 12 different primitive vectors L' of norm m for which the rotation (L, L') is governed by this class. All of these rotations are of the form $(L, \varepsilon L'\varepsilon^{-1})$, where (L, L') is one such rotation and ε runs through all unit quaternions not connected by the equations $\varepsilon' = \pm\varepsilon$. We shall call this set of similar rotations a sheaf of rotations.

By using the consideration presented above, B. A. VENKOV [1] deduced the well-known theorem of GAUSS, which reduces to the following equations $(m>3)$:

$$\text{(IV.2.4)}\quad H_0(m)=\begin{cases}12h(-m) & \text{for } m\equiv 1,2 \pmod 4,\\ 24h'(-m)=8h(-m) & \text{for } m\equiv 3 \pmod 8,\end{cases}$$

where $h(-m)$ is the class number of properly primitive forms of determinant $(-m)$, $h'(-m)$ the class number of improperly primitive forms with the same determinant.

Further, if L and L' are primitive vectors of norm m and r a given odd number, then there exists a primitive quaternion C such that

$$CLC^{-1}=L', \quad b+L=AC, \quad (\operatorname{Norm}(C), r)=1\,.$$

If $m\equiv 1, 2, \pmod 4$, then it is even possible to attain $(\operatorname{Norm}(C), 2r)=1$. The form $(\operatorname{Norm}(A), b, \operatorname{Norm}(C))$ governs the rotation (L, L').

We shall also use Theorem III.1.5 on the composition of forms.

§ 3. Primitive Points in the Fundamental Triangle

We return now to the fundamental spherical triangle Ω and its primitive points, described by primitive vectors L. If $L\in\Omega$ and $\varepsilon\neq\pm 1$, then $\varepsilon L\varepsilon^{-1}\notin\Omega$. In fact, the unit quaternions induce revolutions around the bisectors and reflections in the coordinate planes, and therefore carry L out of Ω.

If $L\in\Omega$, $L_1\in\Omega$, and the rotation (L, L_1) is governed by the form (a, b, c) according to the equations (IV.2.3), then the quaternion A in the equation $b+L=AC$ is uniquely determined up to its sign. Obviously, A is determined up to a unit multiple ε, and if $A^{-1}LA=L_1\in\Omega$, then it follows from

$$\varepsilon_1^{-1}A^{-1}LA\varepsilon_1=\varepsilon_1^{-1}L_1\varepsilon_1\in\Omega$$

that $\varepsilon_1=\pm 1$.

Let $\Gamma_0\subset \mathrm{Sp}_3(1)$ be a region of the type given in Theorem IV.1.1 (and Γ its projection onto $\mathrm{Sp}_3(m)$).

We consider the four-dimensional unit sphere

$$\mathrm{Sp}_4(1)\colon x_1^2+x_2^2+x_3^2+x_4^2=1\,.$$

Its points will be interpreted as quaternions τ with norm one and the points of $\mathrm{Sp}_3(1)$ as vectors $v=\xi i+\eta j+\zeta k$ (quaternions with zero real parts). If $\nu \in \mathrm{Sp}_3(1)$, then $\tau^{-1}\nu\tau \in \mathrm{Sp}_3(1)$. If $\Gamma_0 \subset \mathrm{Sp}_3(1)$ then it is clear that $\tau^{-1}\Gamma_0\tau \subset \mathrm{Sp}_3(1)$. Let ν_0 be any point on $\mathrm{Sp}_3(1)$ and $U(\nu_0, \Gamma_0)$ be the set of points $\tau \in \mathrm{Sp}_4(1)$ such that $\tau^{-1}\nu_0\tau \in \Gamma_0$. This region in $\mathrm{Sp}_4(1)$ decomposes into a finite number of pieces with piecewise smooth boundaries. Further, with the help of the theory of Haar measures on groups (HALMOS [1]) it is easy to prove that

(IV.3.1) $$\frac{\operatorname{meas} U(\nu_0, \Gamma_0)}{\operatorname{meas} \mathrm{Sp}_4(1)} = \frac{\operatorname{meas} \Gamma_0}{4\pi},$$

where "*meas*" denotes the measure on the surface of $\mathrm{Sp}_4(1)$ or $\mathrm{Sp}_3(1)$ accordingly.

For a given region Γ_0 (assumed to be closed and convex) we consider the intersection $\Gamma_0' = \Gamma_0 \cap \Omega_0$. This intersection is also convex in $\mathrm{Sp}_3(1)$, but not necessarily closed.

The quaternion unit ε lies in $\mathrm{Sp}_4(1)$. Let us look at the region $\varepsilon^{-1}\Omega_0\varepsilon \subset \mathrm{Sp}_3(1)$. Twelve such regions form a disjoint covering of the sphere $\mathrm{Sp}_3(1)$. Denote all such regions by $\Omega_0, \Omega_1, \ldots, \Omega_{23}$, and set $\Gamma_k' = \Gamma_0 \cap \Omega_k$. If we prove Theorem IV.1.1 in the case where Γ_0 is replaced by Γ_0', then it is obvious that the theorem holds true for Γ_0. The projections of primitive points belonging to A into Γ_k may be carried over to Ω_0 by means of "integral rotations" with the help of unit quaternions or revolutions. Therefore we shall consider in the following only integral primitive points of Γ', centrally projected from Γ_0' to $\mathrm{Sp}_3(m)$.

For $m > 3$ there are primitive points in Ω; the number of such points is equal to $\dfrac{H_0(m)}{24}$. Let $L^{(0)}$ be one such point.

We consider the rotations $(L^{(0)}, L)$, where L runs through all primitive points on Ω, including $L^{(0)}$. According to the principles developed in § 2 the rotations $(L^{(0)}, L)$ are governed by a class of primitive binary forms. If $m \equiv 1, 2 \pmod 4$, then this class is properly primitive; for $m \equiv 3 \pmod 8$ it is improperly primitive.

Let $(a_1), (a_2), \ldots, (a_h)$ be all of the properly primitive classes $(h = h(-m))$, and for $m \equiv 3 \pmod 8$ denote by $(b_1), \ldots, (b_{h'})$ the improperly primitive classes $(h' = h'(-m))$. To the vector $L^{(0)}$ and the class (a_i) (or (b_i), according to m) there corresponds in Ω one and only one vector $L^{(a_i)}$ – the result of the rotation of $L^{(0)}$ governed by the class (a_i) (or (b_i)).

We handle as an example the easier case: $m \equiv 1, 2 \pmod 4$. Let $q \geqq 3$ be prime with $(-m/q) = +1$. We choose one of the two solutions to the equation $\zeta_0^2 + m \equiv 0 \pmod q$ and construct the properly primitive quadratic form $\varphi = (q, \zeta_0, \varrho)$ with determinant $(-m)$. Denote the class of this form by q. We remark that the classes $1, q, \ldots, q^t$ are different when

(IV.3.2) $$q^t < \sqrt{m}, \quad t < \frac{1}{2}\,\frac{\ln m}{\ln q}.$$

We note also that the class q^ν may be composed with any of the classes a_i or b_i; the class $a_i q^\nu$ will be properly primitive and $b_i q^\nu$ improperly primitive.

Now set

(IV.3.3)
$$s=\left[\frac{1}{2}\frac{\ln m}{\ln q}-1\right].$$

We begin the proof of the Ergodic Theorem IV.2.1. First replace Γ_0 by Γ_0'. The indicator function of the set Γ_0' on $\mathrm{Sp}_3(1)$, that is, $f_{\Gamma_0'}(X)$, will be denoted by $\varphi(X)$. Let $L_0, L_1, \ldots, L_g$ be all primitive vectors belonging to Ω. These will coincide with the vectors $L^{(a_i)}$ and $L^{(b_i)}$ introduced above. We solve the congruence

$$l^2+m\equiv 0\,(\operatorname{mod} q^s),$$

where $l\equiv\zeta_0\,(\operatorname{mod} q)$ and ζ_0 is the quantity fixed above. With this l we write the equations

(IV.3.4)
$$l+L_\alpha=P_\alpha \mathrm{V}_\alpha,$$

where P_α, V_α are integral quaternions and $\operatorname{Norm}(P_\alpha)=q^s$.

In the following $\varepsilon_0, \varepsilon_1, \ldots$ will be small fixed positive numbers, chosen one after another. The number ε_0 is chosen arbitrarily small.

We shall partition the sphere $\mathrm{Sp}_3(1)$ (and project the partition to $\mathrm{Sp}_3(m)$) by means of a lattice of meridians and parallels. In particular, we shall take $z=0$ on $\mathrm{Sp}_3(1)$ as the equator and $y=0$ as the prime meridian. The meridians and the parallels are assumed to be equally spaced.

We consider the lattice generating this partition and choose the subdivisions small enough, so that the measure of each cell on $\mathrm{Sp}_3(1)$ does not exceed $\varepsilon_1=\varepsilon_1(\varepsilon_0)$. To each cell we adjoin its southwest side; the segments of the parallel and the meridian which lie to the bottom and left respectively are adjoined. The north and east sides are rejected. To the cells at the poles are adjoined their west sides; the north and south poles are excluded.

Let $\Lambda_1, \Lambda_2, \ldots, \Lambda_n$ be the intersections of the cells obtained above with Ω_0. Here, $n\leqq c_1$; Λ_i is a convex set on the sphere with piecewise smooth boundary. In each of these sets Λ_i we choose and fix an arbitrary point v_i $(i=1, 2, \ldots, n)$, which we write as a vector in the quaternion algebra.

Let Λ_{i_0} be one such Λ_i. Consider the vectors L with endpoints on the projection of Λ_{i_0} to $\mathrm{Sp}_3(m)$, assuming that the equation (IV.3.4) holds. For the given number k in the formulation of Theorem IV.2.1 we take the number $k_1=kk_2$ and write the equation (IV.3.4) in the form

(IV.3.5)
$$l+L_\alpha=R_{\alpha_1}R_{\alpha_2}\ldots R_{\alpha_{s_1}}W_\alpha,$$

where R_{α_j} are integral quaternions with norm q^{k_1} and $s_1=[s/k_1]$, $\alpha=1, 2, \ldots, h(\Lambda_{i_0})$; $h(\Lambda_{i_0})$ being the corresponding number of primitive vectors.

Let us now determine which $R_{\alpha_1}, R_{\alpha_2}, \ldots,$ we shall choose from each bunch of associated quaternions. We introduce the notation

(IV.3.6) $\quad R_{\alpha_1} \ldots R_{\alpha_m} = T_{\alpha_m}, T_{\alpha_m} \mathrm{Norm}(T_{\alpha_m})^{-\frac{1}{2}} = S_{\alpha_m} \in \mathrm{Sp}_4(1)\,.$

Let T_{α_m} be chosen from the bunch of right associates such that $T_{\alpha_m}^{-1} L_\alpha T_{\alpha_m} \in \Omega$. This defines T_{α_m} in this bunch up to its sign. Thus the R_{α_m} are defined successively. Fix the signs arbitrarily.

We consider now the region $U(v_{i_0}, \Gamma_0') \subset \mathrm{Sp}_4(1)$. Its measure in $\mathrm{Sp}_4(1)$ is expressed by (IV.3.1). If $\Gamma_0' = \Omega_0$, then meas $U(v_{i_0}, \Gamma_0') = (1/24)$ meas $\mathrm{Sp}_4(1)$, and

$$\text{(IV.3.7)} \qquad \frac{\operatorname{meas} U(v_{i_0}, \Gamma_0')}{\operatorname{meas} U(v_{i_0}, \Omega_0)} = \frac{6 \operatorname{meas} \Gamma_0'}{\pi}\,.$$

§ 4. Reduction of the Problem to the Calculation of Probabilities of Large Deviations

Let us consider any one of the equations (IV.3.5) and construct the vectors $S_{\alpha\mu}^{-1} v_{i_0} S_{\alpha\mu}$ $(\mu = 1, 2, \ldots, S_1)$; we seek for the number of times these vectors belong to $\Gamma_0'' \subset \Gamma_0'$, where Γ_0'' is a convex region obtained from by the following: If

$$\frac{1}{\sqrt{m}} T_{\alpha\mu}^{-1} L_\alpha T_{\alpha\mu} \in \Gamma_0'\,,$$

then $S_{\alpha\mu}^{-1} v_{i_0} S_{\alpha\mu} \in \Gamma_0''$.

For each point in Γ_0'' we shall find a point in Γ_0' so that the distance between the two points on $\mathrm{Sp}_3(1)$ is less than ε_2, where ε_2 is the maximal diameter of the lattice cells which we may take to be some small angle. Thus,

$$\text{(IV.4.1)} \qquad \operatorname{meas} \Gamma_0' > \operatorname{meas} \Gamma_0'' > \operatorname{meas} \Gamma_0'(1 - \varepsilon_3)\,.$$

We consider the following question: Which ones among the $S_{\alpha\mu}$ $(\mu = 1, 2, \ldots, s_1)$ satisfy the condition

$$\text{(IV.4.2)} \qquad S_{\alpha\mu} \in U(v_{i_0}, \Gamma_0'')?$$

The corresponding quaternion $T_{\alpha\mu} = S_{\alpha\mu} \mathrm{Norm}(T_{\alpha\mu})^{1/2}$ will be a primitive quaternion of norm $q^{k_1\mu}$.

In the following a special case of the results obtained in the monograph of A. V. MALYSHEV and presented here in the introduction will be important. These results concern the four-dimensional sphere and we will formulate them in terms of quaternions.

Let the region U_0 in $\mathrm{Sp}_4(1)$ be given, bounded by a finite number of convex surfaces, and let v be a large odd number. Expanding $\mathrm{Sp}_4(1)$ by $\sqrt{v}$ we obtain $\mathrm{Sp}_4(v)$ and in $\mathrm{Sp}_4(v)$ the region U as the projection of U_0. Denote by $\mathcal{M}(v)$ the total number of integral primitive points on $\mathrm{Sp}_4(v)$ (primitive quaternions of norm v); by $\mathcal{M}(v, U_0)$ the number of these points for which the projection to $\mathrm{Sp}_4(1)$ lies in U_0; by $\mathcal{M}(v, Q_0)$ the number of primitive quaternions not left divisible by the quaternion Q_0 with $\mathrm{Norm}(Q_0) < \ln v$; by $\mathcal{M}(v, Q_0, U_0)$ the number of such quaternions whose projections to $\mathrm{Sp}_4(1)$ lie in U_0. We then have

Theorem IV.4.1 (A. V. Malyshev):

(IV.4.3) $$\mathcal{M}(v, Q_0, U_0) = \mathcal{M}(v, Q_0)\,\frac{\mathrm{meas}\, U_0}{\mathrm{meas}\,\mathrm{Sp}_4(1)}\left(1 + 0\left(\frac{1}{\ln^2 v}\right)\right).$$

In particular, for $Q_0 = 1$

(IV.4.4) $$\mathcal{M}(v, U_0) = \mathcal{M}(v)\,\frac{\mathrm{meas}\, U_0}{\mathrm{meas}\,\mathrm{Sp}_4(1)}\left(1 + 0\left(\frac{1}{\ln^2 v}\right)\right).$$

Here U_0 is fixed and $v \to \infty$. $\mathcal{M}(v, Q_0) = \mathcal{M}_0(v)$ depends on Q_0. In (IV.4.2) we shall take $\mu = 1$. If the divisibility conditions (IV.3.5) are not imposed on the quaternion $T_{\alpha 1} = R_{\alpha 1}$, then the number of $T_{\alpha 1}$ of norm q^{k_1} satisfying $S_{\alpha_1} \in U(v_{i_0}, \Omega_0)$ will be, according to Theorem IV.4.1,

(IV.4.5) $$\mathcal{M}(q^{k_1}, U(v_{i_0}, \Omega_0)) = \mathcal{M}(q^{k_1}) \cdot \frac{1}{24}\left(1 + \frac{\theta c_2}{\ln^2 q^{k_1}}\right)$$

(in the following, θ will represent a number with $|\theta| \leqq 1$, and fixed).

The number of $T_{\alpha 1}$ such that $S_{\alpha 1} \in U(v_{i_0}, \Gamma_0'')$ is given by

(IV.4.6) $$\mathcal{M}(q^{k_1}, U(v_{i_0}, \Gamma_0'')) = \mathcal{M}(q^{k_1})\,\frac{1}{24}\,u\left(1 + \frac{\theta c_2}{\ln^2 q^{k_1}}\right),$$

where we set

(IV.4.7) $$u = \frac{\mathrm{meas}\, U(v_{i_0}, \Gamma_0'')}{\mathrm{meas}\, U(v_{i_0}, \Omega_0)} = \frac{6\,\mathrm{meas}\,\Gamma_0''}{\pi}.$$

If $T_{\alpha j}$ is fixed, then $T_{\alpha, j+1} = T_{\alpha j} R_{\alpha, j+1}$; thus if $T_{\alpha j}$ is right divisible by $\bar{Q}_0$ with $\mathrm{Norm}(Q_0) = q$, then $R_{\alpha, j+1}$ is primitive and not left divisible by Q_0. Further, for fixed $T_{\alpha j}$, $R_{\alpha, j+1}$ is determined among its right associates up to its sign (and thus $T_{\alpha, j+1}$ is determined). If

(IV.4.8) $$S_{\alpha, j+1}^{-1}\, v_{i_0}\, S_{\alpha, j+1} \in \Gamma_0',$$

then

(IV.4.9) $$R_{\alpha, j+1}^{-1}\, S_{\alpha j}^{-1}\, v_{i_0}\, S_{\alpha j}\, R_{\alpha, j+1} \in \Gamma_0''.$$

The number of $R_{\alpha, j+1}$ satisfying the condition (IV.4.9) is, according to the fundamental Theorem IV.4.1, given by

(IV.4.10) $$\mathcal{M}_0(q^{k_1})\,\frac{u}{24}\left(1 + \frac{\theta c_2}{k_1^2 \ln^2 q}\right).$$

If for given α and arbitrary $j + 1 \leqq s_1 + 1$ in the equations (IV.3.5), $T_{\alpha, j+1}$ and $S_{\alpha, j+1} = T_{\alpha, j+1}\,\mathrm{Norm}(T_{\alpha, j+1})^{1/2}$ are determined so that the condition (IV.4.8) is fulfilled, then we shall say that "the event U occurs in row α at the $(j+1)$-th place".

We consider the products $R_{\alpha 1}, R_{\alpha 2}, \ldots, R_{\alpha s_1}$ which may occur (but do not necessarily occur) in the first part of (IV.3.5).

With the help of (IV.4.6) and (IV.4.10) we shall give an asymptotic expression for the number of products

(IV.4.11) $$R_{\alpha 1} \dots R_{\alpha s_1}$$

in which the event U takes place r times. Let this occur at the places $j_1, \dots, j_r$ and not at the other places.

For all products (IV.4.11) we have

(IV.4.12) $$\frac{1}{24^{s_1}} \mathcal{M}(q^{k_1}) (\mathcal{M}_0(q^{k_1}))^{s_1-1} = \mathcal{W}(q^{k_1 s_1}),$$

where the right side is the number of right non-associated primitive quaternions of norm $q^{k_1 s_1}$. From (IV.4.6) and (IV.4.10) we deduce that for the products of the form (IV.4.11), where the event U occurs exactly r times and at the places $j_1, \dots, j_r$, we have the expression

(IV.4.13) $$\frac{1}{24^{s_1}} \mathcal{M}(q^{k_1 j_1}) \mathcal{M}_0(q^{k_1(j_2-j_1)}) \dots \mathcal{M}_0(q^{k_1(j_r-j_{r-1})}) \times \\ \times \mathcal{M}_0(q^{k_1(s_1-j_r)}) u^r (1-u)^{s_1-r} \prod_{\nu=1}^{s_1} \left(1+\frac{\theta_\nu c_2}{k_1^2 \ln^2 q}\right),$$

where $|\theta_\nu| \leq 1$. This number is equal to

(IV.4.14) $$u^r (1-u)^{s_1-r} \mathcal{W}(q^{k_1 s_1}) \prod_{\nu=1}^{s_1} \left(1+\frac{\theta_\nu c_2}{k_1^2 \ln^2 q}\right).$$

The number of products (IV.4.11) for which the event U occur exactly r times will be

(IV.4.15) $$C_{s_1}^r u^r (1-u)^{s_1-r} \mathcal{W}(q^{k_1 s_1}) \Pi(\theta, k_1)$$

where $\Pi(\theta, k_1)$ lies between the greatest and least of the products having the form

$$\prod_{\nu=1}^{s_1} \left(1+\frac{\theta_\nu c_2}{k_1^2 \ln^2 q}\right)$$

and occurring in the expressions (IV.4.14). Obviously we have

(IV.4.16) $$\exp(-\eta(k_1) s_1) < \Pi(\theta, k_1) < \exp(\eta(k_1) s_1),$$

where $\eta(k_1) \to 0$ as $k_1 \to \infty$.

The factor

$$C_{s_1}^r u^r (1-u)^{s_1-r}$$

has in part a propabilistic meaning; it is the probability of occurrence of an event exactly r times in a Bernoulli scheme with s_1 independent trials if the probability of the event in one trial is equal to u.

§ 5. Calculation of Probabilities of Large Deviations. An Application of Theorem III.2.1

Let $\zeta_1 > 0$ be a given small number (in the following $\zeta_1, \zeta_2, \dots$ are small positive numbers, consecutively chosen). It is known that if X is the random number of appearances of the event in the scheme given above, then

$$E(X)=us_1, \quad \sigma(X)=\sqrt{u(1-u)\,s_1}\,.$$

We introduce now the compound event M_{ζ_1}, described by

(IV.5.1) $$s_1(u-\zeta_1)\leqq X\leqq s_1(u+\zeta_1),$$

and its negation $\bar{M}_{\zeta_1}$. For fixed $\zeta_1>0$ and u and for sufficiently large s_1 we have according to Theorem III.2.1

(IV.5.2) $$P(M_{\zeta_2})\leqq \exp(-\zeta_2 s_1),$$

where P denotes probability.

Recall now the equations (IV.3.5). If their number is

(IV.5.3) $$h(\Lambda_{i_0})<\frac{h(-m)}{\ln^2 m},$$

then we shall not consider these equations, and we select the another index i_0, one for which the inequality (IV.5.3) is violated (obviously, for the fixed partition and sufficiently large m such an index may be found). Therefore we shall assume that

(IV.5.4) $$h(\Lambda_{i_0})\geqq\frac{h(-m)}{\ln^2 m}.$$

Choose from the $h(\Lambda_{i_0})$ equations (IV.3.5) any g of them with

(IV.5.5) $$g\geqq\frac{h(\Lambda_{i_0})}{\ln^2 m}$$

and write down the corresponding products (IV.4.11). If the event $\bar{M}$ happens in all such products, then the number of different products will not be greater than

(IV.5.6) $$\mathscr{W}(q^{k_1 s_1})\,P(\bar{M})\,\Pi(\theta,k_1)\leqq\mathscr{W}(q^{k_1 s_1})\exp(-\zeta_2 s_1)\exp(\eta(k_1)s_1)\,.$$

For sufficiently large k_1 we see that

$$\eta(k_1)s_1<\tfrac{1}{2}\zeta_2 s_1,$$

and the number of products does not exceed

(IV.5.7) $$\mathscr{W}(q^{k_1 s_1})\exp(-\tfrac{1}{2}\zeta_2 s_1)<q^{k_1 s_1(1-\zeta_3(k_1))},$$

where $\zeta_3(k_1)>0$ depends on k_1. Further, according to (IV.5.4) and (IV.5.5)

(IV.5.8) $$g\geqq\frac{h(\Lambda_{i_0})}{\ln^2 m}\geqq\frac{h(-m)}{\ln^4 m}>c(\eta)m^{\frac{1}{2}-\eta}$$

for arbitrary $\eta>0$ (a theorem of K. L. SIEGEL, see (IV.1.1)).

Let s_2 be chosen such that

$$q^{s_2}\leqq m^{\frac{1}{2}+\zeta_4}, \quad q^{s_2+1}>m^{\frac{1}{2}+\zeta_4}, \quad \zeta_4>0\,.$$

In the g equations of (IV.3.5) which we have selected, we choose now $l_1\equiv l \pmod{q^{s_2}}$ and form the new equations

(IV.5.9) $$l+L_\alpha=R_{\alpha 1}\dots R_{\alpha s_1}R_{\alpha,s_1+1}\dots R_{\alpha s_2}W'_\alpha\,.$$

If the event $\bar{M}_{\zeta_1}$ occurs in all products $R_{\alpha 1} \dots R_{\alpha s_1}$, then according to (IV.5.7) the number of different products $R_{\alpha 1} \dots R_{\alpha s_1}$ will not exceed

(IV.5.10) $$2q^{k_1(s_2-s_1)}q^{k_1 s_1(1-\zeta_3(k_1))} < m^{\frac{1}{2}(1-\zeta_4(k))}$$

for sufficiently small ζ_4. At the same time, the number of distinct L_α in the equations (IV.5.9) is approximated by (IV.5.8). We can now choose ζ_4 so small, that (IV.5.10) is satisfied; in (IV.5.8) we choose $\eta < \frac{1}{2}\zeta_4(h_1)$. Then a contradiction is obtained to the key Theorem III.2.1. Hence we conclude that it is impossible to select g equations with g satisfying the condition (IV.5.5) and such that the event $\bar{M}_{\zeta_1}$ occurs in all products (for sufficiently large m). Consequently, in all of the equations (IV.3.5) the event M_{ζ_1} occurs, with the possible exception of not more than

$$\frac{h(\Lambda_{i_0})}{\ln^2 m}$$

of them, or

$$h(\Lambda_{i_0}) < \frac{h(-m)}{\ln^2 m}.$$

Collecting the equations (IV.3.5) for all i_0 we conclude that in all equations

(IV.5.11) $$l + L_\alpha = R_{\alpha 1} \dots R_{\alpha s_1} V_\alpha,$$

where $L_\alpha \in \Omega$, the event M_{ζ_1} is fulfilled in all products $R_{\alpha 1} \dots R_{\alpha s_1}$ with the possible exception of

$$O\left(\frac{h(-m)}{\ln^2 m}\right)$$

such products.

§ 6. Completion of the Proof to the Ergodic Theorem IV.2.1

Occurrence of the event M_{ζ_1} means that for given i_0, X being the number of α for which

(IV.6.1) $$S_{\alpha j}^{-1} v_{i_0} S_{\alpha j} \in \Gamma_0'',$$

the inequalities (IV.5.1) are fulfilled. Here $\Gamma_0'' \subset \Gamma_0'$; the boundaries of these regions are close to each other, and their areas satisfy the inequality (IV.4.1). If the condition (IV.6.1) is fulfilled, we have

(IV.6.2) $$\frac{1}{\sqrt{m}} T_{\alpha j}^{-1} L_\alpha T_{\alpha j} \in \Gamma_0'.$$

Therefore, the number of places j in the row α for which (IV.6.2) is valid will be

(IV.6.3) $$r_\alpha \geqq s_1(u - \zeta_1).$$

We now desire to obtain an upper bound for r_α. For this we substitute for Γ_0' the doubly-connected region $\Omega_0 - \Gamma_0'$. To this region we apply the same reasoning, and for all α, possibly excepting $O\left(\frac{h(-m)}{\ln^2 m}\right)$ and sufficiently large m we get: the number r'_α of those j for which

(IV.6.4) $$\frac{1}{\sqrt{m}} T_{\alpha j}^{-1} L_\alpha T_{\alpha j} \in \Omega_0 - \Gamma'_0$$

is subject to the inequality

(IV.6.5) $$r'_\alpha \geqq s_1(1-u-\zeta_1).$$

Then, obviously,

(IV.6.6) $$r_\alpha = s_1 - r'_\alpha \leqq s_1 - s_1(1-u-\zeta_1) = s_1(u+\zeta_1).$$

The relations (IV.6.3) and (IV.6.2) are valid for all α with $L_\alpha \in \Omega$, possibly excepting

$$O\left(\frac{h(-m)}{\ln^2 m}\right)$$

of them.

This completes one step in the proof of the Ergodic Theorem IV.2.1.

In the equations (IV.5.11) we examine the $R_{\alpha j}$'s, whose norm is q^{k_1}, and the initial quaternion of norm $k = k_1/k_2$. Denoting those quaternions by $Q_{\alpha\mu}$ we obtain from (IV.5.11))

(IV.6.7) $$l + L_\alpha = Q_{\alpha 1} \dots Q_{\alpha s_3} V'_\alpha,$$

where $s_3 = k_2 s_1$. Here $Q_{\alpha j}$ is determined up to its sign in its bunch of associates by the condition that

(IV.6.8) $$\frac{1}{\sqrt{m}} (Q_{\alpha 1} \dots Q_{\alpha j})^{-1} L_\alpha Q_{\alpha 1} \dots Q_{\alpha j} \in \Omega_0.$$

Let μ be an integer with $0 \leqq \mu \leqq k_1 - 1$. We introduce the following notation:

$$R_{\alpha 1} = \prod_{j=1}^{\mu} Q_{\alpha j},\ R_{\alpha 2} = \prod_{j=\mu+1}^{\mu+k_1} Q_{\alpha j}, \dots, R_{\alpha \nu} = \prod_{j=(\nu-2)k_1+\mu+1}^{(\nu-1)k_1+\mu} Q_{\alpha j} \quad \text{for } \nu \geqq 2$$

for values of ν satisfying

$$(\nu - 1)k_1 + \mu \leqq s_3.$$

Writing (IV.6.7) in the form

(IV.6.9) $$l + L_\alpha = R_{\alpha 1} \dots R_{\alpha s} Z_\alpha,$$

and beginning with $j = 2$, we can carry out for the factors $T_{\alpha j} = R_{\alpha 1} \dots R_{\alpha j}$, the same consideration as previously. Setting $\mu = 0, 1, \dots, k_1 - 1$ we get that, with the possible exception of

$$k \cdot O\left(\frac{h(-m)}{\ln^2 m}\right),$$

the number of α for which the condition (IV.6.2) is fulfilled will be bounded by the inequalities

(IV.6.10) $$s_1(1-u-\zeta_1) \geqq r_\alpha \geqq s_1(u-\zeta_1).$$

As $m \to \infty$ the number of exceptions will be $O\left(\frac{h(-m)}{\ln^2 m}\right)$, since k_1 is a fixed number.

We remark that $u=\frac{6\operatorname{meas}\Gamma_0''}{\pi}$; in view of (IV.4.1) u is bounded by

(IV.6.11) $$u_1 \geqq u \geqq u_1(1-\varepsilon_3), \quad u_1 = 6/\pi \operatorname{meas}\Gamma_0'.$$

Therefore we obtain from the preceding:

Let

(IV.6.12) $$l+L_\alpha = Q_{\alpha 1} \dots Q_{\alpha s_3} V_\alpha',$$

where s_3 is any number with $s_3 > c_0 \ln m$. We set $T_{\alpha j} = Q_{\alpha 1} \dots Q_{\alpha j}$ and note the number r_α of occurrences, for $j=1, 2, \dots, s_3$, of the event

(IV.6.13) $$\frac{1}{\sqrt{m}} T_{\alpha j}^{-1} L_\alpha T_{\alpha j} \in \Gamma_0'.$$

For a fixed small ζ_5 we shall have:

(IV.6.14) $$s_3(u+\zeta_5) \geqq r_\alpha > s_3(u-\zeta_5)$$

for sufficiently large m and all α except possibly

$$O\left(\frac{h(-m)}{\ln^2 m}\right)$$

of the α's.

We shall investigate what happens when we substitute $s > s_3$ for s_3 and $l' \equiv l \pmod{q^{s_3}}$, where $l'^2 + m \equiv 0 \pmod{q^s}$, writing equation (IV.6.7) with the number s. From § 3 we see that if $j = h(-m)$ (the class number of properly primitive forms) or in general a multiple of the number h_0 of the class q in the group H, then $T_{\alpha j}^{-1} L_\alpha T_{\alpha j} = L_\alpha$ (according to the above conditions on the choice of $T_{\alpha j}$ in its corresponding bunch). For $j > h_0$, $T_{\alpha j}$ begins to repeat itself periodically. Therefore it is sufficient to consider the values s for which

$$c_0 \ln m \leqq s \leqq h_0.$$

From the above argument we see that c_0 may be chosen arbitrarily small but fixed; we shall take such a c_0 and set $s = [c_0 \ln m]$. If

$$l + L_\alpha = R_{\alpha 1} \dots R_{\alpha s} Z_\alpha, \quad T_{\alpha j} = R_{\alpha 1} \dots R_{\alpha j},$$

then for fixed $j = j_0$ the set of $L_\alpha \in \Omega$ will coincide with the set of $T_{\alpha j}^{-1} L_\alpha T_{\alpha j}$. This result follows from Theorem III.1.5 on the composition of classes.

We consider now the case

$$h_0 \geqq s \geqq (\ln m)^{3/2}$$

(if it is possible, that is, if q has large enough bounds). The mapping

(IV.6.15) $$T_{\alpha j}^{-1} L_\alpha T_{\alpha j} = L_{\alpha'}$$

takes the set $U(L_\alpha \in \Omega)$ into itself for fixed j.

Let $j = 0, s_0, 2s_0, \dots, [s/s_0]\, s_0$. We shall replace L_α by $L_{\alpha'}$ according to the formula (IV.6.15), write the corresponding equations of type

(IV.5.7), and denote by $\bar{M}_{\zeta_5}(j_0)=\bar{M}_{j_0}$ the exceptional sets of values, those for which the event

(IV.6.16) $$s_0(u+\zeta_5)\geqq r_\alpha > s_0(u-\zeta_5)$$

does not occur. Let the number $\mathcal{M}(N)$ of elements be $\mathcal{M}(N)=h_1 = O\left(\frac{h(-m)}{\ln^2 m}\right)$. The sets N may intersect each other for different j_0; we shall count the indices entering into $L_{\alpha'}$ with repetitions. Considering now equations of the form

(IV.6.17) $$l'+L_{\alpha'}=Q_{\alpha' 1}\dots Q_{\alpha' s_0}W_{\alpha'},$$

obtained from equations of the form (IV.6.7) by replacing L_α by $L_{\alpha'}$ according to (IV.6.15), we shall call the index α "poor", if for $j_0=0$, s_0, $2s_0$, ..., $[s/s_0]\,s_0$ in the corresponding equations (IV.6.17) no less than

(IV.6.18) $$\zeta_6\cdot s/s_0$$

indices $\alpha'\in N_{j_0}$ appear. The total number of "poor" indices α for all α will be smaller than or equal to $\frac{s}{s_0}h_1$. Therefore, if h_2 is the number of "poor" indices, then we get

(IV.6.19) $$h_2\zeta_6\frac{s}{s_0}\leqq\frac{s}{s_0}h_1,$$
$$h_2\leqq\frac{h_1}{\zeta_6}=O\left(\frac{h(-m)}{\ln^2 m}\right).$$

For "good" indices we have obviously

(IV.6.20) $$s(u+\zeta_5)+s_0\geqq r_\alpha > s(u-\zeta_5)-s_0;$$

hence, taking into account

$$\frac{s}{s_0}=O((\ln m)^{-\frac{1}{4}}),$$

we obtain

(IV.6.21) $$s(u+2\zeta_5)\geqq r_\alpha > s(u-2\zeta_5)$$

for sufficiently large s.

We consider now the case

$$s_0\leqq s<(\ln m)^{3/2}.$$

For this case let $j_0=0, 1, \dots, s-s_0$. Writing the equations of the form (IV.6.17) for α', derived from (IV.6.15), we see that $\alpha'\in\bar{M}_{j_0}$. The total number does not exceed

$$sh_1=O\left(\ln^{3/2}m\cdot\frac{h(-m)}{\ln^2 m}\right),$$

that is, their number has the order $O\left(\frac{h(-m)}{\sqrt{\ln m}}\right)$.

For the remaining "good" α we have the expression (IV.6.21). The equivalent Theorem IV.2.1 is proved, concerning flows of projected primitive points $L_\alpha \in \Omega$ defined with the help of the quaternion transformation

(IV.6.22)
$$L' = Q^{-1} L Q,$$

where $\mathrm{Norm}(Q) = q^k$; Q a primitive quaternion uniquely determined up to its sign, and thus (IV.6.22) is determined uniquely and carries L into $L' \in \Omega$. It remains only to translate the operation of flows into the language of orthogonal matrices.

§ 7. Orthogonal Matrices. A Mixing Theorem. The Asymptotic Distribution of Primitive Points on the Sphere

As is known, the general form of orthogonal transformations of three-dimensional space with determinant $+1$ is given by the formula

$$X' = A^{-1} X A$$

where A is a quaternion with $\mathrm{Norm}(A) > 0$. If T is the transformation matrix, then we have

(IV.7.1)
$$X' = A^{-1} X A = T \begin{Vmatrix} x \\ y \\ z \end{Vmatrix},$$

where the notation is obvious. Let $T = \begin{Vmatrix} \frac{a_{ij}}{q^k} \end{Vmatrix}$; $\det(T) = +1$; not all $\frac{a_{ij}}{q^k}$ reducible. Then $q^k T$ will be an integral matrix, and the quaternion $q^k A^{-1} X A$ is integral for all properly integral X (whose real part may differ from zero), and, for any X with $\mathrm{Norm}(X)$ relatively prime to q^k, primitive. Hence it follows from elementary properties of the arithmetic of quaternions that $A = \alpha Q$, where Q is an integral quaternion with norm q^k and α a real number. Thus, the transformation $X' = T \begin{Vmatrix} x \\ y \\ z \end{Vmatrix}$ will correspond uniquely (up to the sign of Q) to the mapping $X' = Q^{-1} X Q$, the quaternion Q having norm q^k. Conversely, if such a transformation is given, then it will be orthogonal. In (IV.7.1) T satisfies $\det(T) = +1$, and we conclude from the equation $X' = Q X Q q^{-k}$ that $T = \begin{Vmatrix} \frac{a_{ij}}{q^k} \end{Vmatrix}$ with a_{ij} integral, even if Q is an improperly integral quaternion, and not all of the fractions $\frac{a_{ij}}{q^k}$ are reducible.

Thus, the basic operation T of our flow may be expressed as a unimodular orthogonal transformation with matrix T.

We turn now to Theorem IV.2.2 on mixing. We shall deduce it from the Ergodic Theorem IV.2.1, maintaining the same partition of $\mathrm{Sp}_3(1)$ and the same notation as in the proceeding paragraphs.

Let N be any set of indices α with $L_\alpha \in \Omega$, their number being

(IV.7.2) $$\mathcal{M}(N) > \varepsilon_0 h(-m).$$

Let $\Lambda_1, \Lambda_2, \ldots, \Lambda_n$, $n \leqq c_1$ be the intersections of the partition cells with Ω_0 (see § 3). We introduce the notation

$$N_\mu = N \cap \Lambda_\mu \qquad (\mu = 1, 2, \ldots, n).$$

Consider $L_\alpha \in N_\mu$ and take into account only those values μ for wich

(IV.7.3) $$\mathcal{M}(N_\mu) > \frac{h(-m)}{\ln^2 m}.$$

We form the equation

(IV.7.4) $$l + L_\alpha = R_{\alpha 1} R_{\alpha 2} \ldots R_{\alpha s_0} V_\alpha, \quad \mathrm{Norm}(R_{\alpha j}) = q^k.$$

The discussion will be conducted as in LINNIK [6] (p. 211).

Separate the second indices j of the quaternions $R_{\alpha j}$ into two types: Type 1, those indices j for which the number λ_j of first indices α fulfilling the condition

(IV.7.5) $$S_{\alpha j}^{-1} v_\mu S_{\alpha j} \in \Gamma_0''$$

(see § 4) satisfies the inequality

(IV.7.6) $$\lambda_j \geqq (1 - \zeta_7) u \mathcal{M}(N_\mu);$$

Type 2, those indices j for which the inequality is not satisfied. If there are less than $s_1 = \zeta_8 s_0$ of the indices of type 2, then Λ_μ will be called a "good" region; if not, then Λ_μ will be called a "poor" region. First it is necessary to prove for sufficiently large m, and for μ satisfying the condition (IV.7.3), that the region Λ_μ is "good".

Let Λ_μ be a "poor" region. Then among the indices of the quaternions $R_{\alpha j}$ appearing on the right side of equation (IV.7.4), there are not less than $s_1 \geqq \zeta_8 s_0$ second indices j_k for which the number of first indices α satisfying the condition

(IV.7.5) $$S_{\alpha j}^{-1} v_\mu S_{\alpha j} \in \Gamma_0''$$

does not exceed $(1 - \zeta_7) \cdot u \cdot \mathcal{M}(N_\mu)$. Let these s_1 indices be $j_1, j_2, \ldots, j_{s_1}$. We form the matrix of the corresponding vectors $L_{\alpha j}$:

$$\| L_{\alpha j_k} \|.$$

Let h be the number of its rows in which the condition (IV.7.5) is fulfilled not more than $(1 - \zeta_7/2) u s_1$ times. Then we have

$$(1 - \zeta_7) u \mathcal{M}(N_\mu) s_1 \geqq (\mathcal{M}(N_\mu) - h)(1 - \zeta_7/2) u s_1,$$

hence

$$h \geqq \frac{\zeta_7}{2 - \zeta_7} \mathcal{M}(N_\mu) = \zeta_9 \mathcal{M}(N_\mu).$$

Thus, among the indices α with $K_\alpha \in N_\mu$ we find at least $h_2 > \zeta_9 \mathcal{M}(N_\mu)$ of them such that for s_3 fixed second indices $j_1, j_2, \ldots, j_{s_3}$, $s_3 \geqq s_1/2$, the number r_α of $T_{\alpha j_\beta}$ for which

(IV.7.7) $$T_{\alpha j_\beta}^{-1} v_\mu T_{\alpha j_\beta} \in \Gamma_0''$$

satisfies the inequality

(IV.7.8) $$r_\alpha < (1-\zeta_{10}) u s_3 .$$

Here ζ_{10} may be chosen to be independent of the maximal diameter of Λ_i on $\mathrm{Sp}_3(1)$.

Putting $R'_{\alpha\beta} = T_{\alpha j_{\beta-1}}^{-1} T_{\alpha j_\beta}$, we consider for the h_2 indices chosen above the equation

(IV.7.9) $$l + L_\alpha = R'_{\alpha 1} R'_{\alpha 2} \ldots R'_{\alpha s_3} V_\alpha .$$

We have

(IV.7.10) $$\prod_{l=1}^{v} R'_{\alpha l} = T_{\alpha j_v} .$$

Treating the transformation $T_{\alpha j_v}^{-1} L_\alpha T_{\alpha j_v}$ exactly as we have treated the transformation $T_{\alpha j}^{-1} L_\alpha T_{\alpha j}$, we obtain an exact analogue to the Ergodic Theorem IV.2.1 (in the case of all indices sufficiently dense subsequences $j_1, \ldots, j_{s_3}$ are chosen one after another; everything follows as above). According to this theorem, the number of α satisfying the condition (IV.7.7) is of the order $O\left(\dfrac{h(-m)}{\ln^2 m}\right)$, which contradicts the condition $h_2 > \zeta_9 \mathcal{M}(N_\mu)$. This shows that Λ_μ must be a "good" region.

We haved proved that every region Λ_μ satisfying condition (IV.7.3) is "good". The number of indices of type 2 in each of these regions is less than $s_1 \leqq \zeta_8 s_0$, and in all of them less then $n s_1 < c_1 s_1 \leqq c_1 \zeta_8 s_0 = \zeta_{11} s_0$. Therefore in all regions Λ_μ satisfying (IV.7.3) with the corresponding equations (IV.7.4) there are more than $(1-\zeta_{11}) s_0$ indices of type 1. Producing the corresponding transformations

(IV.7.11) $$T_{\alpha j}^{-1} L_\alpha T_{\alpha j} ,$$

where j remains fixed and is of the type 1, α takes all values on such that $L_\alpha \in \Omega$, and taking into account that different L_α produce different inequalities (IV.7.3), and that the N_μ do not satisfy the condition (IV.7.3), it is easy to see that for all $l \leqq s_0$, except possibly indices $l \leqq \zeta_{11} s_0$,

(IV.7.12) $$\mathcal{M}(T^l N \cap \Gamma_0') > (1-2\zeta_1) \frac{6}{\pi} \operatorname{meas} \Gamma_0' ,$$

where T is the flow operation of (IV.7.10).

Replacing Γ_0' by $\Omega_0 - \Gamma_0'$, we find that for $l \leqq s_0$, with the possible exception of $l \leqq 2\zeta_{11} s_0$ indices, the expression (IV.7.11) and the inequality

(IV.7.13) $$\mathcal{M}(T^l N \cap \Gamma_0') < (1+2\zeta_7)\, 6/\pi \operatorname{meas} \Gamma_0'$$

are fulfilled. In the case s_0 it is possible to take any $s \geqq s_0$ and proceed according to § 6.

Formulating the flow operations in the language of orthogonal matrices, we get Theorem IV.2.2 on mixing.

Theorem IV.1.1 turns out to be a direct corollary of Theorem IV.2.2 on mixing. If we take $N = A$ (the set of all primitive $L_\alpha \in \Omega$), then, for all l, $T^l A = A$, so that from the mixing theorem the asymptotic uniform distribution of L_α on Ω follows, and also in general on $\mathrm{Sp}_3(m)$, which is the content of Theorem IV.1.1.

If we assume $m \equiv 1, 2 \pmod 4$, then all of the governing classes of forms are properly primitive, and we can prove a generalization of Theorem IV.1.1. Let $\sigma < H$ be a subgroup of the group of classes with bounded index $n \leqq c_3$ and G_i $(i = 0, 1, \ldots, n-1)$ the cosets.

We take any $L_0 \in \Omega$ and let each class G_i operate on it, transforming it in Ω. Thus we obtain the set of primitive points $H(G_i)$. It is always possible to choose k sufficiently large and satisfying the condition $q^k \in G_0$.

Determining the flow with the help of q^k, we see that it does not take $L_\alpha \in H(G_i)$ out of this set, and in general, transforms $H(G_i)$ into itself. Applying Theorem IV.2.2 on mixing, we obtain the generalization of Theorem IV.4.1:

Theorem IV.4.1'. *The set of primitive points $H(G_i)$ is asymptotically uniformly distributed on $\mathrm{Sp}_3(m)$ as $m \to \infty$ (in the sense given in the formulation of Theorem IV.1.1).*

§ 8. Supplementary Remarks

We have considered until now only primitive points on the sphere $\mathrm{Sp}_3(m)$ and proved the asymptotic relation (IV.1.2). The relation (IV.1.3), that is, the set of all integral points on the sphere considered asymptotically, is easily deduced from the above. If the point (x, y, z) is not primitive, then we have $(x, y, z) = \delta(x', y', z')$, where (x', y', z') is primitive. The integral points can be classified according to their divisors δ. Fix a small $\varepsilon_1 > 0$. If the divisor satisfies $\delta^2 \geqq m^{\varepsilon_1}$, then the quantity of all such points will be of the order $0(m^{\frac{1}{2}-\varepsilon_1+\varepsilon})$ for every $\varepsilon > 0$, and we may neglect them. If $\delta^2 < m^{\varepsilon_1}$, then we consider the primitive points on $\mathrm{Sp}_3(m/\delta^2)$ and deduce for them the corresponding form (IV.1.2). Here it is necessary to sum over such relations for all δ^2/m, $\delta^2 < m^{\varepsilon_1}$, which reduces to the formula (IV.1.3).

An interesting problem turns out to be that of the order of the term $\kappa_0(\Gamma_0, q, m)$ for given Γ_0, q and as $m \to \infty$. Specifying the previous discussions and making better use of the theorem on probabilities of large deviations I.3.1, it is possible to attain

$$\kappa_0(\Gamma_0, q, m) = O\left(\frac{1}{(\ln m)^\alpha}\right), \tag{IV.8.1}$$

where α is a positive number. However, to advance further with the given methods is difficult.

The existence of the auxiliary number q also turns out to be a deficiency of the method. This may be avoided if some hypotheses on the zeroes of L-series are proved, which turn out to be weak corollaries of a generalized Riemann hypothesis. Under theses circumstances, as will be shown in Chapter IX, it is possible to deduce Theorem IV.1.1 by comparatively elementary means, but with a poorer remainder than (IV.8.1).

A significant improvement of the remainder in the theorem on the sphere is hardly possible without completely new methods, in particular, without deeper penetration into the nature of Kloosterman sums (see LINNIK [9]).

We remark further, that the problems considered here on the asymptoticity of integral points on $Sp_3(m)$ (but not on their ergodic behavior) have been widely generalized in a series of papers by MALYSHEV, where asymptotic distributions of integral points on ellipsoids of sufficiently arbitrary type are given.

Chapter V

Flows of Primitive Points on a Hyperboloid of Two Sheets. Asymptoticity of Reduced Binary Forms in Connection with Lobachevskian Geometry

§ 1. Formulation of the Problem

In this chapter we shall consider the same questions as in Chapter IV for flows of primitive points on half of a hyperboloid of two sheets. This problem is difficult in that the surface is not compact and contains infinitely many primitive points.

These difficulties are overcome by treating, instead of the complete surface, a (compactified) fundamental region on the hyperboloid. The area measure on this region will be introduced by means of Lobachevskian geometry. The arithmetic of quaternions is replaced here by the arithmetic of integral matrices $\begin{pmatrix} \alpha & \beta \\ \gamma & \delta \end{pmatrix}$; the basic ideas of this arithmetic are found in Chapter II, § 2.

We examine half of the two-sheeted hyperboloid H in cartesian coordinates (a, b, c):

(V.1.1) $$ac - b^2 = D; \quad D > 0; \quad a > 0.$$

Integral points (a, b, c) are considered to be images of positive binary quadratic forms

(V.1.2) $$(a, b, c) = ax^2 + 2bxy + cy^2.$$

Among the primitive integral points we select those where $(a, 2b, c) = 1$. These will be called admissible; they correspond to the properly primitive forms among (V.1.2).

The conditions on the form (V.1.2) for it to be reduced in the sense of Lagrange are:

(V.1.3) $$|2b| < a < c, \quad \text{or} \quad 0 \leqq 2b < a = c, \quad \text{or} \quad 0 < 2b = a < c.$$

Introducing the new coordinates $x_1 = a/\sqrt{D}$, $x_2 = c/\sqrt{D}$, $x_3 = b/\sqrt{D}$, we obtain the normed hyperboloid H_0

(V.1.4) $$x_1 x_2 - x_3^2 = 1, \quad x_1 > 0.$$

This hyperboloid may be interpreted as a Lobachevskian plane (see B. A. VENKOV [3]). We give a short resumé of this interpretation. The straight lines are represented by hyperbolas—sections of H_0 cut out by the planes $a_1x_1+a_2x_2+a_3x_3=0$. The angles are computed with the help of the form $F=x_1x_2-\frac{1}{4}x_3^2$ reciprocal to the left side of (V.1.4).

If the straight lines

$$a_1x_1+a_2x_2+a_3x_3=0\,; \quad b_1x_1+b_2x_2+b_3x_3=0 \tag{V.1.5}$$

are given, with the coefficients normed such that

$$F(a_1,a_2,a_3)=F(b_1,b_2,b_3)=-1\,,$$

then the angle φ between them is given by

$$\cos\varphi=\pm\left(\tfrac{1}{2}(a_1b_2+a_2b_1)-\tfrac{1}{4}a_3b_3\right). \tag{V.1.6}$$

The Lobachevskian area of an arbitrary finitely connected figure, bounded by piecewise smooth curves, possesses a suitable interpretation (VENKOV [3]).

Let S_0 be such a figure in H_0. We construct the cone with vertex (0, 0, 0), based on this figure. Its Euclidean volume will be the Lobachevskian area of S_0. Here, the Lobachevskian constant is $k=\sqrt{\frac{2}{3}}$, so that the area of a triangle with defect δ is $\frac{2}{3}\delta$.

The points on H satisfying the reduction conditions (V.1.3) are represented by the triangle Δ_0 on the hyperboloid H_0. This triangle is bounded by the Lobachevskian lines

$$x_2-x_1=0\,; \quad x_1-2x_3=0\,; \quad x_1+2x_3=0\,. \tag{V.1.7}$$

Its angles are respectively equal to $\pi/3$, $\pi/3$, and 0, and the area is

$$A(\Delta_0)=\tfrac{2}{9}\pi\,. \tag{V.1.8}$$

We shall call Δ_0 the fundamental triangle.

Let $K_1 \geqq 1$ be an arbitrary constant. The line $x_2-K_1x_1=0$ cuts a quadrangle $G_0(K_1)$ off of Δ_0.

We return now to the integral points on H.

Points satisfying the reduction condition (V.1.3) are called fundamental. Admissible and fundamental points correspond to properly primitive reduced forms. The number of such points on H will be $h(-D)$ – the class number of properly primitive forms with determinant $(-D)$. It is known that $h(-D)$ satisfies the asymptotic expression

$$\ln h(-D) \sim \tfrac{1}{2}\ln D \quad \text{as} \quad D\to\infty \tag{V.1.9}$$

(see (3.2.2)).

Let Σ be a convex figure bounded by piecewise smooth curves on the hyperboloid H, such that its image Σ_0 in H_0 lies in the fundamental triangle Δ_0. If this image coincides with Δ_0, then the figure is designated by Δ. Let $H(\Sigma)$ be the number of admissible (and automatically fundamental) points on this figure; for instance, $H(\Delta)=h(-D)$. We shall be interested in the ratio $\dfrac{H(\Sigma)}{H(\Delta)}$ for large D.

§ 2. Formulation of the Basic Theorems

Here we formulate basic theorems of the same character as those for the sphere $Sp_3(m)$ in Chapter IV. The first three of them carry an asymptotic-geometric character, the remaining ones being of ergodic type.

Theorem V.2.1: *Let D be an odd number, $p \geqq 3$ an arbitrary prime with $(-D/p) = +1$, $K_1 \geqq 1$ a constant as large as desired, and Σ_0 a figure in H_0 whose image in H is Σ. Assume that Σ_0 is contained in the quadrangle $G_0(K_1)$. Then the following asymptotic relation is valid:*

$$\frac{H(\Sigma)}{H(\Delta)} = \frac{\Lambda(\Sigma_0)}{\Lambda(\Delta_0)}(1+\eta(p, K_1, D)), \tag{V.2.1}$$

where $\Lambda(\Sigma_0)$ is the Lobachevskian area of Σ_0; $\Lambda(\Delta_0)$ that of Δ_0; and $\eta(p, K_1, D) \to 0$ for fixed p, K_1 as $D \to \infty$.

Since $H(\Delta) = h(-D)$, $\Lambda(\Delta_0) = \frac{2}{9}\pi$, we get

$$H(\Sigma) = \frac{9}{2\pi} h(-D)\,\Lambda(\Sigma_0)\,(1+\eta(p, K_1, D)). \tag{V.2.2}$$

Using the asymptotic expressions (V.2.1) and (V.2.2) and by means of elementary calculation of areas in the Lobachevskian plane theorems directly connected with the theory of L-series of Dirichlet may be deduced.

Denote by $h(-D, \alpha\sqrt{D})$ the number of properly primitive reduced forms (a, b, c) for which $a \leqq \alpha\sqrt{D}$. Then we have Theorem V.2.2:

Theorem V.2.2: *If $\alpha \leqq 1$, we have*

$$h(-D, \alpha\sqrt{D}) = \frac{3\alpha}{\pi} h(-D)\,(1+\eta(p, \alpha, D)). \tag{V.2.3}$$

For $1 \leqq \alpha \leqq \sqrt{\frac{4}{3}}$, we have

$$h(-D, \alpha\sqrt{D}) = f(\alpha)\, h(-D)\,(1+\eta(p, D)), \tag{V.2.4}$$

where

$$f(\alpha) = \frac{6}{\pi} \arcsin \sqrt{1-\frac{1}{\alpha^2}} + \frac{3\alpha}{\pi}\left(1-2\sqrt{1-\frac{1}{\alpha^2}}\right). \tag{V.2.5}$$

For $\alpha \geqq \sqrt{\frac{4}{3}}$ we have $h(-D, \alpha\sqrt{D}) = h(-D)$ (trivial result).

In (V.2.3) and (V.2.4), $\eta(p, \alpha, D) \to 0$ for fixed p, α as $D \to \infty$, and $\eta(p, D) \to 0$ for fixed p as $D \to \infty$.

It is possible to formulate part of Theorem V.2.2 in terms of Dirichlet series.

Let $-D < 0$ be a fundamental discriminant and $X(n) = (-D/n)$ be the corresponding Dirichlet character. We consider the Dirichlet series

$$\zeta(s)\, L(s, X) = \sum_{n=1}^{\infty} a_n n^{-s} \qquad (s > 1).$$

Theorem V.2.3: *Let $\varepsilon > 0$ be an arbitrarily small fixed number. Then*

$$\sum_{n \leqq \varepsilon \sqrt{D}} a_n = \varepsilon \sqrt{D}\, L(1, X)\,(1 + \eta_2(p, \varepsilon, D)), \tag{V.2.6}$$

where $\eta_2(p, \varepsilon, D) \to 0$ for fixed p, ε as $D \to \infty$.

We observe that Theorem V.2.3 would be a direct consequence of the Riemann hypothesis for the series $L(s, X)$ or the weaker, at the present time unproved density hypothesis. For details on this see the following paragraph.

We turn now to theorems of ergodic character (including a theorem on mixing).

If two fundamental points $(a) = (a_0, b_0, c_0)$ and $(b) = (a'_0, b'_0, c'_0)$ are given, then the Gaussian composition of the classes of these forms will give us a third point $(a)(b) = (a''_0, b''_0, c''_0)$. Let $p \geqq 3$ be a fixed prime such that $(-D/p) = +1$. We form any two (mutually reciprocal) forms $(p, \pm\xi, n)$ of determinant $-D$ and denote their class by (p). The fundamental point $(p) = (a'_0, b'_0, c'_0)$ corresponds to this class. If $(a) = (a_0, b_0, c_0)$ is an arbitrary fundamental point, then we form the gaussian composition $(T) = (a''_0, b''_0, c''_0)$ and say that (p) generates a flow of the classes of forms $(a) \to (a)(p)$, being given by the transformation to (T) through the composition of the classes of forms with (p). We denote the result of the composition (ap) by $(a)T$; let $(a)T^l$ designate the fundamental point corresponding to the class of forms $(a)(p)^l$. Let $\Omega \subset \Delta_0$ be a simply connected region of Δ_0 with piecewise smooth boundary, and $\Lambda(\Omega)$ its Lobachevskian area. Set $f_\Omega(P) = 1$ if $P \in \Omega$, $f_\Omega(P) = 0$ if $P \notin \Omega$.

We are interested in the behavior of the ergodic mean of $f_\Omega((a)T^l)$.

Theorem V.2.4:

$$\frac{1}{s}\left[f_\Omega((a)) + f_\Omega((a)T) + \cdots + f_\Omega((a)T^{s-1})\right] = \Lambda(\Omega)\,(1 + o(1)), \tag{V.2.7}$$

if $s > c_0 \ln D$ ($c_i > 0$ are constants) for $D \to \infty$, and for all classes (a) with the possible exception of $o(h(-D))$ of them, where $h(-D)$ is the total class number (the number of elements in the group of classes of H).

Theorem V.2.5: *The expression* (V.2.7) *remains valid for fixed r and $c_0 = c_0(r)$ if we substitute for the given fundamental operation T the operation T^r.*

From the ergodic theorems it is possible to go over to theorems of the mixing type. Let N be any set of classes (a); $\mathcal{M}(N)$ the number of elements of N. Let $\mathcal{M}(N) > \varepsilon_0 h(-D)$, where $\varepsilon_0 > 0$ is an arbitrarily small constant. Let $\mathcal{M}(NT^l, \Omega)$ denote the number of points of N lying in Ω after the transformation T^l. Let $l = 0, 1, 2, \ldots, s \geqq c_2 \ln D$. Then for all indices l, with the possible exception of $0(\ln D)$ such indices, the following theorem is valid.

Theorem V.2.6 *(mixing theorem):*

$$\mathcal{M}(NT^l, \Omega) = \mathcal{M}(N)\,\Lambda(\Omega)\,(1+o(1)) \quad as \quad D \to \infty\,. \tag{V.2.8}$$

In this chapter we shall only prove Theorems V.2.1, V.2.2 and V.2.3, referring the reader for the demonstrations of the remaining theorems to their analogies in Chapter IV.

We consider points (a, b, c) with arbitrary real coordinates lying on the hyperboloid H (V.1.1). The number D will be assumed integral and odd. To the point (a, b, c) will correspond a positive quadratic form

$$\varphi(x, y) = ax^2 + 2bxy + cy^2 \tag{V.2.9}$$

and a matrix with trace zero

$$L = \begin{pmatrix} b & -a \\ c & -b \end{pmatrix}. \tag{V.2.10}$$

These three objects will always correspond to each other and we assume them to be given simultaneously.

As was stated in § 2, Chapter II, if we pass from the algebra of matrices to the algebra of hermitions $\mathscr{E}$, then the matrix

$$L = \begin{pmatrix} b & -a \\ c & -b \end{pmatrix}$$

goes over into the hermition

$$L = \frac{a+c}{2}\, i_1 + b i_2 + \frac{a-c}{2}\, i_3\,.$$

Introducing the substituion

$$x = \frac{a+c}{2\sqrt{D}}, \quad y = \frac{b}{\sqrt{D}}, \quad z = \frac{a-c}{2\sqrt{D}}, \tag{V.2.11}$$

we obtain from our hyperboloid H (C.1.1) a new hyperboloid H_1:

$$x^2 - y^2 - z^2 = 1\,; \quad x + z > 0\,. \tag{V.2.12}$$

If we set $L = x i_1 + y i_2 + z i_3$, then the automorphisms of the form $\varphi = x^2 - y^2 - z^2$, having determinant $+1$ and arbitrary real coefficients, may be expressed by the formula

$$L' = \varepsilon L \varepsilon^{-1}, \tag{V.2.13}$$

where $\varepsilon = \zeta + \mu_1 i_1 + \mu_2 i_2 + \mu_3 i_3$ is an arbitrary hermition with Norm$(\varepsilon) = 1$. This follows from the general form of such automorphisms (see BACHMANN).

On the new hyperboloid H_1 the points are assigned Weierstrass coordinates (z, y, x) (see, for example, USPENSKY). Here the group G_1 of motions of the first type on the Lobachevskian plane, carrying each figure into a congruent one, coincides with the group of automorphisms (V.2.12)

$$V=\begin{pmatrix}\alpha & \beta & \gamma\\ \alpha' & \beta' & \gamma'\\ \alpha'' & \beta'' & \gamma''\end{pmatrix}$$

under the conditions $\det(V)=+1, \alpha>0$ (see USPENSKY, p. 122). We prove now the following lemma.

Lemma V.2.1: *An automorphism V under the conditions stated above is given by the formula* (V.2.13)

$$L'=\varepsilon L\varepsilon^{-1} \quad \textit{if and only if} \quad \operatorname{Norm}(\varepsilon)=+1\,.$$

For the proof set $L'=x' i_1+y' i_2+z' i_3$ and compute $\varepsilon L\bar{\varepsilon}$. We obtain for $\operatorname{Norm}(\varepsilon)=+1$, $\varepsilon^{-1}=\bar{\varepsilon}$:

$$\begin{aligned}
x' &= x(\zeta^2+\mu_1^2+\mu_2^2+\mu_3^2)+y(-2\zeta\mu_3-2\mu_1\mu_2)+z(2\zeta\mu_2-2\mu_1\mu_3)\,,\\
y' &= x(-2\zeta\mu_3+2\mu_1\mu_2)+y(\zeta^2-\mu_1^2-\mu_2^2+\mu_3^2)+z(2\zeta\mu_1-2\mu_2\mu_3)\,,\\
z' &= x(2\zeta\mu_2+2\mu_1\mu_3)+y(-2\zeta\mu_1-2\mu_2\mu_3)+z(\zeta^2-\mu_1^2+\mu_2^2-\mu_3^2)\,.
\end{aligned}$$

Thus if $\operatorname{Norm}(\varepsilon)=+1$, then the condition $\alpha>0$ is satisfied. If $\operatorname{Norm}(\varepsilon)=-1$, then $L'=-\varepsilon L\bar{\varepsilon}$ and $\alpha<0$. Since any automorphism with determinant $+1$ is given by the formula (V.2.13) with $\operatorname{Norm}(\varepsilon)=\pm1$, everything is proved.

Returning to the algebra of matrices and the former hyperboloid H_0 (V.1.4), we set

$$L=\begin{pmatrix}x_3 & -x_1\\ x_2 & -x_3\end{pmatrix}$$

corresponding to (V.2.10). From the statement just proved it follows that if A runs through all matrices with $\det(A)=+1$ and

$$L'=\begin{pmatrix}x_3' & -x_1'\\ x_2' & -x_3'\end{pmatrix},$$

then the transformation

(V.2.14) $$L'=ALA^{-1}$$

produces the whole group G_1 of Lobachevskian motions of the first type.

In the group of unimodular matrices

$$\begin{pmatrix}\alpha & \beta\\ \gamma & \delta\end{pmatrix}$$

there exists a left and right invariant measure. The differential of the invariant measure has the form

$$\frac{d\alpha\, d\beta\, d\gamma}{|\alpha|}$$

for elements with $\alpha \neq 0$, and an analogous form for the remaining points (see, for example, CHEBOTAREV, p. 353).

We denote by $d\mu$ the differential of the invariant measure. Let Ω be a measurable region on the group of unimodular matrices. We consider all matrices

$$A=\begin{pmatrix}\alpha & \beta\\ \gamma & \delta\end{pmatrix}$$

for which

$$0<u_1 \leqq \det(A) \leqq u_2 .$$

The matrix

$$\frac{A}{+\sqrt{\det(A)}}$$

is unimodular. Let us prove the following lemma.

Lemma V.2.2: *For any positive* u_1 *and* $u_2 > u_1$ *we have*

$$\text{(V.2.15)} \qquad \iiint\limits_{A' \in \Omega} d\mu = \frac{2}{u_2^2 - u_1^2} \int \cdots \int\limits_{G} d\alpha\, d\beta\, d\gamma\, d\delta ,$$

where G is the region of all $(\alpha, \beta, \gamma, \delta)$ *for which*

$$\text{(V.2.16)} \qquad u_1 \leqq \alpha\delta - \beta\gamma \leqq u_2 ; \quad A' = \frac{1}{\sqrt{\alpha\delta-\beta\gamma}}\begin{pmatrix}\alpha & \beta\\ \gamma & \delta\end{pmatrix} \in \Omega .$$

For the proof split Ω into disjoint sets such that on each of the sets one of the coordinates $(\alpha, \beta, \gamma, \delta)$ is never zero. Without loss of generality we may assume $\alpha \neq 0$ on all of Ω. Introducing the new coordinates

$$u = \alpha\delta - \beta\gamma, \quad \alpha' = \frac{\alpha}{\sqrt{u}}, \quad \beta' = \frac{\beta}{\sqrt{u}}, \quad \gamma' = \frac{\gamma}{\sqrt{u}},$$

it is easy to calculate that

$$\frac{\partial(\alpha, \beta, \gamma, \delta)}{\partial(\alpha', \beta', \gamma', \delta')} = \frac{u}{|\alpha'|} .$$

Then the integral on the right side of (V.2.15) becomes

$$\frac{2}{u_2^2 - u_1^2} \int\limits_{u_1}^{u_2} u\, du \int \cdots \int\limits_{A' \in \Omega} \frac{d\alpha'\, d\beta'\, d\gamma'}{|\alpha'|} = \int \cdots \int\limits_{A' \in \Omega} d\mu ,$$

which was to be shown. Formula (V.2.15) will be useful in the sequel.

Let Γ be a region in the hyperboloid H_0 bounded by convex piecewise smooth curves, and x_1, x_2, x_3 be a point on H_0 with the corresponding matrix

$$L = \begin{pmatrix} x_3 & -x_1 \\ x_2 & -x_3 \end{pmatrix} .$$

Consider the set of all motions carrying (x_1, x_2, x_3) to $\Gamma: L' = ALA^{-1}$, where A is a unimodular matrix. The set Ω of the corresponding matrices A will have measure $\mu(\Omega)$; it is not hard to show, using formula (V.2.15), that this quantity is finite. This measure is not changed by a shift of the region Γ by L_0 or by the replacement of the point (x_1, x_2, x_3) by any other one.

By the same token $\mu(\Omega)$ induces on H_0 a measure $\mu'(\Gamma)$, invariant under the group of Lobachevskian motions of the first type.

Using the theorem on the uniqueness of Haar measure up to a multiplicative constant (Halmos [1]), we find that our measure $\mu(\Omega)$ on the group of unimodular matrices is proportional to the Lobachevskian area $\Lambda(\Gamma)$ on H_0. This is necessary for the following results. Formula (V.2.15) gives us a suitable expression, proportional to Lobachevskian area on H_0.

Now we shall collect data on integral matrices A which produce rotations

$$L' = ALA^{-1} \tag{V.2.17}$$

of an integral vector matrix

$$L = \begin{pmatrix} b & -a \\ c & -b \end{pmatrix}$$

with odd $\det(L) = ac - b^2 = D > 0$ into the same type of integral vector matrix. We consider vector matrices L corresponding to the admissible points (a, b, c), that is, properly primitive forms $\varphi(x, y) = ax^2 + 2bxy + cy^2$. With the methods in § 1 of Chapter III and by elementary results extended from quaternions to the algebra $\mathscr{E}$, we obtain the following important lemma.

Lemma V.2.3: *Let*

$$L = \begin{pmatrix} b & -a \\ c & -b \end{pmatrix}, \quad L' = \begin{pmatrix} b' & -a' \\ c' & -b' \end{pmatrix}$$

be two admissible vector matrices with the same odd norm D: $D = -L^2 = -L'^2$. Then there exists a primitive vector matrix Q such that

$$\det(Q) > 0; \quad (\det(Q), 2D) = 1; \quad QLQ^{-1} = L'. \tag{V.2.18}$$

Further calculations follow exactly the same lines as in § 1 of Chapter III. From (V.2.18) we see that $QL = L'Q$, so that L belongs to the right principal ray (mod Q); hence it follows that there exists an integer l with

$$l + L = PQ, \tag{V.2.19}$$

where P is an integral primitive matrix.

Since $\det(l + L) = l^2 + D > 0$ and $\det(Q) > 0$, we have $\det(P) > 0$.

Construct the quadratic form

$$\Psi(\xi,\eta)=\det(P)\xi^2+2l\xi\eta+\det(Q)\eta^2.$$

This form turns out to be properly primitive. We say that the form $\Psi(\xi,\eta)$ governs the (non-euclidean) rotation (L, L'). We have

$$\Psi(\xi,\eta)=(\bar{P}\xi+Q\eta)(P\xi+\bar{Q}\eta).$$

If we perform the unimodular integral substitution

$$\begin{aligned}\bar{P}_1&=\bar{P}x_1+Qx_2\\ Q_1&=\bar{P}x_3+Qx_4,\end{aligned}\qquad \begin{vmatrix}x_1 & x_2\\ x_3 & x_4\end{vmatrix}=1,$$

then the form $\Psi(\xi,\eta)$ is transformed into an equivalent form which governs the same rotation.

Further we see that inequivalent forms govern different rotations $L\to L'$ and $L\to L''$, $L'\neq L''$. One and the same properly primitive form $\Psi(\xi,\eta)$ may correspond to infinitely many sets of equations of the form

(V.2.20) $$l+L=P\bar{\varepsilon}\cdot\varepsilon Q,$$

where $\varepsilon=\begin{pmatrix}\alpha & \beta\\ \gamma & \delta\end{pmatrix}$ is unimodular. If $QLQ^{-1}=L'$, then $\varepsilon QL(\varepsilon Q)^{-1}=\varepsilon L'\bar{\varepsilon}$. The vector matrix

$$L'=\begin{pmatrix}l' & -a'\\ c' & -l'\end{pmatrix}$$

corresponds to the properly primitive form

$$\varphi'(x,y)=a'x^2+2l'xy+c'y^2,$$

and the vector matrix $\varepsilon L'\bar{\varepsilon}$, according to Lemma 2.2.1, will correspond to the equivalent form $\varphi(x,y)\,S$, where

$$S=\varepsilon^T=\begin{pmatrix}\alpha & \gamma\\ \beta & \delta\end{pmatrix}.$$

In this case, according to equation (V.2.19), each form of a class of properly primitive forms transforms the vector matrix L into a vector matrix L' corresponding to an equivalent form $\varphi'(x,y)$.

Thus, if we choose L and L' from the fundamental region (V.1.3), that is, the points (a,b,c) and (a',b',c') are chosen to be fundamental and admissible, then for given L the rotations $L\to L'$ will be governed by different classes of forms. We obtain the following lemma.

Lemma V.2.3: *If $L_1, L_2, \ldots, L_h$ $(h=h(-D))$ are all of the vector matrices corresponding to fundamental and admissible points (a,b,c) on the hyperboloid H, then each rotation $L_i\to L_j$ is governed by some class of properly primitive forms, and for given L_i and different L_j these classes are different.*

Therefore if the matrices L such that $L^2=-D$ are divided into classes of equivalent matrices, assuming equivalence when the condition

$L' = \varepsilon L \varepsilon^{-1}$ is valid for a unimodular matrix ε, then the number of classes of such matrices is quite simply connected with the number of ideal classes $h(\sqrt{-D})$. This is a special case of a known algebraic theorem which we shall discuss in Chapter VII.

§ 3. Formulation of the Basic Lemma

The basic lemma has the same character and plays an equivalent role for rotations of the hyperboloid as Theorem III.2.1 for rotations of the sphere. Let $p \geqq 3$ be a fixed prime number such that $(-D/p) = +1$. Then the congruence $l^2 + D \equiv 0 (\operatorname{mod} p^s)$ is solvable for any $s > 0$. From Chapter II § 2 we know that there exist $U(p^s) = p^{s-1}(p+1)$ left non-associated primitive matrices $\prod^{(i)}$ with $\det(\prod^{(i)}) = p^s$. For every L_i $(i = 1, 2, \ldots, h)$ in Lemma V.2.3 we construct the matrix $l + L_i$, where l is one of the two solutions of the congruence $l^2 + D \equiv 0 (\operatorname{mod} p^s)$, fixed for all matrices L_i. From the unique factorization theorem up to units for matrices we obtain the h equations

(V.3.1) $$l + L_i = \prod\nolimits_i X_i \qquad (i = 1, 2, \ldots, h).$$

Here the X_i are integral matrices and each $\prod_i$ has the form $\prod^{(j)} \varepsilon_{ij}$, where $\prod^{(j)}$ is one of the matrices given above. In the following let $\eta_1, \eta_2, \ldots$ be small positive constants to be determined successively and $c_1, c_2, \ldots$ constants greater then 1.

Consider any h_1 of the equations in (V.3.1) with

$$h_1 \geqq D^{\frac{1}{2} - \eta_1}, \quad \eta_1 < 0.1 .$$

We number them from 1 to h_1. The number l in (V.3.1) is assumed to be dependent on s and taken to satisfy

(V.3.2) $$0 < l < p^s .$$

Basic Lemma: *Choose η_1 sufficiently small and set*

(V.3.3) $$\eta_2 = 10 \eta_1 .$$

Let s be selected such that

(V.3.4) $$D^{\frac{1}{2} + \eta_2} < p^s < p D^{\frac{1}{2} + \eta_2} .$$

Form the h_1 equations

(V.3.5) $$l + L_i = \prod\nolimits_i X_i \qquad (i = 1, 2, \ldots, h_1).$$

We have

(V.3.6) $$\det\left(\prod\nolimits_i\right) = p^s ; \quad \prod\nolimits_i = \prod\nolimits^{(j)} \varepsilon_{ij} .$$

Then the number of distinct $\prod^{(j)}$ in the equations (V.3.5) *is not less than $D^{\frac{1}{2} - \eta_3}$, where*

(V.3.7) $$\eta_3 = 12 \eta_2 .$$

(Here the ε_{ij} are unit matrices.)

The proof of this basic lemma requires a series of auxiliary arguments and proceeds in analogy to the theory presented in Chapters II and III. The most difficult point turns out to be as in Chapter III the approximation to the number of representations of a binary form in ternary (here the form $x^2-y^2-z^2$ under a few restrictions). In addition there are specific difficulties connected with the indefiniteness of the quaternary form $\det(A)=\alpha\delta-\beta\gamma$.

By the expression $A=0(\varphi(B_1,\dots,B_n))$, where A is a matrix, φ a positive function, and $B_1,\dots,B_n$ run through some infinite set of values, we shall mean that the coefficients of the matrix A divided by $\varphi(B_1,\dots,B_n)$ are universally bounded by an absolute constant.

Lemma V.3.1: *Suppose that for $D\to\infty$ the two vector matrices L and L' satisfy*

(V.3.8) $$L^2=L'^2=-D\,;\quad L=0(D^{\frac{1}{2}+\tau_0})\,;\quad L'=0(D^{\frac{1}{2}+\tau_0})\,,$$

$\tau_0>0$ being constant. Further let

$$\det(U)=u\geqq 1\,;\quad ULU^{-1}=L' \quad or \quad U^{-1}LU=L'\,,$$

for the matrices L, L'; U not necessarily integral. Then

(V.3.9) $$U=0(\sqrt{u}\,D^{\tau_0})\,.$$

Proof: Set

$$U=\begin{pmatrix}\alpha & \beta\\ \gamma & \delta\end{pmatrix};\quad L=\begin{pmatrix}b & -a\\ c & -b\end{pmatrix};\quad L'=\begin{pmatrix}b' & -a'\\ c' & -b'\end{pmatrix}.$$

We may assume $a>0$, $c>0$, so that L corresponds to the positive quadratic form $ax^2+2bxy+cy^2$; if this is not true, then we may replace L by $-L$ and L' by $-L'$. Further, since $\det(U)\geqq 1$, we see from Lemma II.2.1 that $a'>0$, $c'>0$. By Lemma II.2.1 we have for $\varphi(x,y)=ax^2+ {}+2bxy+cy^2$

$$\varphi'(x,y)=a'x^2+2b'xy+c'y^2=\varphi(x,y)\frac{1}{\sqrt{u}}\begin{pmatrix}\alpha & \gamma\\ \beta & \delta\end{pmatrix}.$$

In addition to this,

$$a'=a\left(\frac{\alpha}{\sqrt{u}}\right)^2+2b\frac{\alpha}{\sqrt{u}}\frac{\beta}{\sqrt{u}}+c\left(\frac{\beta}{\sqrt{u}}\right)^2$$

and

$$aa'=\left(a\frac{\alpha}{\sqrt{u}}+b\frac{\beta}{\sqrt{u}}\right)^2+D\left(\frac{\beta}{\sqrt{u}}\right)^2.$$

In view of (V.3.8), $aa'=0(D^{1+2\tau_0})$. Hence

$$D\left(\frac{\beta}{\sqrt{u}}\right)^2=0(D^{1+2\tau_0})\,;\quad \beta=0(\sqrt{u}\,D^{\tau_0})\,.$$

Analogously we have

$$ca'=\left(c\frac{\beta}{\sqrt{u}}+b\frac{\alpha}{\sqrt{u}}\right)^2+D\left(\frac{\alpha}{\sqrt{u}}\right)^2=0(D^{1+2\tau_0})\,;\quad \alpha=0(\sqrt{u}\,D^{\tau_0})\,.$$

In exactly the same way we can show that $\gamma = 0(\sqrt{u}\, D^{\tau_0})$, $\delta = 0(\sqrt{u}\, D^{\tau_0})$. Further, $U^{-1}LU = \bar{U}L\bar{U}^{-1}$. This proves our lemma. We now turn our attention to the existence of points with $L^2 = -D < 0$ and $\det(U) > 0$.

§ 4. Continuation of the Proof of the Basic Lemma

We shall continue the proof of the basic lemma.

Lemma V.4.1: *For any $M > 1$ and arbitrarily small $\zeta > 0$, the number of fundamental admissible vector matrices*

$$L = \begin{pmatrix} b & -a \\ c & -b \end{pmatrix}$$

for which $c > MD^{\frac{1}{2}}$ is of the order $0(M^{-1}D^{\frac{1}{2}+\zeta})$.

Proof: From the reduction conditions (V.1.3) we get

$$D = ac - b^2 \geqq a\left(c - \frac{a}{2}\right),$$

hence

$$a \leqq \frac{D}{c - a/2} \leqq \frac{2D}{c}.$$

If $c > MD^{\frac{1}{2}}$, then $a \leqq 2M^{-1}D^{\frac{1}{2}}$.

The problem is reduced to the calculation of the number of positive properly primitive reduced forms for which $a \leqq 2M^{-1}D^{\frac{1}{2}}$. According to § 2 of Chapter III, the number of such forms with fixed a is of the order $0(D^{\eta}(D, a)^{\frac{1}{2}})$; in view of this the quantity in which we are interested can be approximated by

$$2M^{-1}D^{\frac{1}{2}} \sum_{\substack{r|D \\ r \leqq 2M^{-1}D^{\frac{1}{2}}}} r^{-\frac{1}{2}}\, 0(D^{\eta}),$$

Since

$$\sum_{r|D} r^{-\frac{1}{2}} \leqq \sum_{r|D} 1 = 0(D^{\eta}),$$

we obtain the required estimate with $\zeta = 2\eta$.

We return now to the equations (V.3.5). Their number is $h_1 \geqq D^{\frac{1}{2}-\eta_1}$. We set $M = D^{\eta_1 + 2\zeta}$ in Lemma V.4.1; the number of the vector matrices L for which $c > D^{\frac{1}{2}+\eta_1+2\zeta}$ will be of the order $0(D^{\frac{1}{2}-\eta_1-\zeta})$. For the remaining ones we get, obviously, in view of (V.1.3), $L = 0(D^{\frac{1}{2}+\eta_1+2\zeta})$.

We see that, for sufficiently large D, among the h_1 equations (V.3.5) there will be not less than $h_2 \geqq \frac{1}{2}h_1 \geqq \frac{1}{2}D^{\frac{1}{2}-\eta_1}$ of them such that the corresponding L_i satisfy

$$L_i = 0(D^{\frac{1}{2}+\eta_1+\zeta_1}),$$

where $\zeta_1 > 0$ is arbitrarily small. Take now instead of (V.3.5) the above mentioned h_2 equations

$$l + L_i = \prod_i X_i, \tag{V.4.1}$$

numbered in succession: $i = 1, 2, \ldots, h_2$, in which the matrices $\prod_i$ are either equal or non-associated. We have

(V.4.2) $$L_i = 0(D^{\frac{1}{2}+\eta_1+\zeta_1}).$$

Further let $P_1 P_2, \ldots, P_w$ be all of the different right non-associated $\prod_i$ appearing in the equations (V.4.1). We collect the equations (V.4.1) into w systems, each system having the same P_j, and obtain equations of the form

$$\begin{aligned} l + L_i &= P_1 Q_i \quad (i = 1, 2, \ldots, a_1), \\ l + L_i &= P_2 Q_i \quad (i = a_1 + 1, \ldots, a_1 + a_2), \\ &\vdots \\ l + L_i &= P_w Q_i \quad (i = a_1 + \cdots + a_{w-1} + 1, \ldots, a_1 + \cdots + a_w), \end{aligned}$$

in which

$$a_1 + \cdots + a_w = h_2 .$$

Now we make the assumption that

(V.4.3) $$w < D^{\frac{1}{2}-\eta_3}.$$

The reduction of this assumption to a contradiction will prove the basic lemma.

First we make a remark on the selection of the non-associated P_j in our equations. If $P_j^{-1} L_i P_j = L'_i$, then replacing P_j by $P_j \varepsilon$ results in a transformation giving $\varepsilon^{-1} L'_i \varepsilon$. From the preceding we know that there exist exactly two unimodular matrices $\pm\varepsilon$ for which, if the admissible L'_i is given, the vector matrix $\varepsilon^{-1} L_i \varepsilon$ will be admissible and fundamental. We shall then assume, that the multipliers P_j are selected among their right associates in accordance with this. However, it is possible that the relation

(V.4.4) $$L'_i = 0(D^{\frac{1}{2}+\eta_i+\zeta_i})$$

is not fulfilled for certain L'_i. According to lemma V.4.1, and because the L'_i are fundamental and admissible, the number of them for which (V.4.4) is violated is of the order $0(D^{\frac{1}{2}-\eta_1-\frac{\zeta_1}{2}})$. For sufficiently large D this number is less than $h^2/2$. We discard now all equations $l + L_i = P_j Q_i$ which lead to such L'_i. Then we get $h^3 \geqq \frac{1}{2} h_2 \geqq \frac{1}{4} D^{\frac{1}{2}-\eta_1}$ equations for which (V.4.4) is fulfilled. To simplify the notation, we shall assume that this was done at the beginning, and instead of h_3 we shall write $h_2 \geqq \frac{1}{4} D^{\frac{1}{2}-\eta_1}$.

Now we obtain together with each equation $l + L_i = P_j Q_i$ the equation $l + L'_i = Q_i P_j$, where L'_i is fundamental and admissible and satisfies (V.4.4). Taking conjugates, we get the equations $l - L'_i = \bar{P}_j \bar{Q}_i$. Note that the vector matrix $(-L'_i)$ is not fundamental. Further, for each system

$$G_k : l + L_i = P_k Q_i \quad (i = a_1 + \cdots + a_{k-1} + 1, \ldots, a_1 + \cdots + a_k)$$

we have the corresponding system

$$\bar{G}_k: l - L'_i = \bar{P}_k \bar{Q}_i \quad (i = a_1 + \cdots + a_{k-1} + 1, \ldots, a_1 + \cdots + a_k).$$

Consider now the rotations of the form $L_{i_1} \to (-L'_{i_2})$, where $L_{i_1} \in G_k$, $-L'_{i_2} \in G_k$. Such rotations cannot be governed by positive properly primitive classes. We show that they are governed by negative properly primitive classes. Let

$$L_{i_1} = \begin{pmatrix} b & -a \\ c & -b \end{pmatrix}; \quad a > 0, c > 0; \quad i_2 = \begin{pmatrix} 1 & 0 \\ 0 & -1 \end{pmatrix}.$$

We have

$$i_2 L_{i_1} i_2^{-1} = \begin{pmatrix} b & a \\ -c & -b \end{pmatrix} = -L''_{i_1}, \quad \text{where} \quad L''_{i_1} = \begin{pmatrix} -b & -a \\ c & b \end{pmatrix}.$$

Obviously, L''_{i_1} is an admissible fundamental vector matrix (corresponding to the form $(a, -b, c)$ inverse to (a, b, c)). Here the rotation $L''_{i_1} \to L'_{i_2}$ is governed by a positive properly primitive class according to the equations

$$b_1 + L''_{i_1} = A_1 C_1; \quad C_1 L''_{i_1} C_1^{-1} = L'_{i_2}.$$

This means that

$$C_1(-L''_{i_1}) C_1^{-1} = -L'_{i_2}.$$

Hence we deduce the equations

$$\bar{A}_1(-L''_{i_1}) \bar{A}_1^{-1} = -L'_{i_2}; \quad b_1 - L''_{i_1} = \bar{C}_1 \bar{A}_1;$$

further,

$$b_1 + L_{i_1} = i_2 \bar{C}_1 \cdot \bar{A}_1 i_2$$

and

$$b_1 - L'_{i_2} = \bar{A}_1 i_2 \cdot i_2 \bar{C}_1 \quad (\text{because } i_2^2 = 1).$$

In addition,

$$\det(\bar{A}_1 i_1) = -\det(\bar{A}_1) < 0; \quad \det(i_1 \bar{C}_1) = -\det(\bar{C}_1) < 0,$$

and we see that the rotation $L_{i_1} \to (-L'_{i_2})$ is governed by the negative properly primitive class

$$(-\det(A_1), b_1, -\det(C_1)).$$

From Lemma V.2.3 and these considerations we obtain the following lemma.

Lemma V.4.2: *If* $L_1, L_2, \ldots, L_h$ $(h = h(-D))$ *are admissible and fundamental vector matrices, then each rotation* $L_i \to (-L_j)$ *is governed by a negative properly primitive class with the representative* $(-a_1, b_1, -c_1)$ *according to the equations*

$$b + L_i = A_1 C_1; \quad C_1 L_i C_1^{-1} = -L_j$$

and

$$\det(A_1) = -a_1; \quad \det(C_1) = -c_1.$$

§ 5. Study of Rotations

Consider all of the rotations $L_i \to (-L_j)$ in Lemma V.4.2. From each of the corresponding governing negative properly primitive classes we select a representative $(-a_1, b_1, -c_1)$ such that $a_1 \geqq c_1 \geqq 2b_1$.

We divide the set of all negative forms satisfying the condition above, but not necessarily properly primitive, into two types: the system of senior forms M and the system of junior forms m. By definition

$$\varphi \in M \quad \text{if} \quad 1 \leqq c_1 \leqq D^{\frac{1}{2}-\tau_1};$$

$$\varphi \in m \quad \text{if} \quad D^{\frac{1}{2}-\tau_1} < c_1 \leqq \sqrt{\frac{4}{3} D},$$

where $\tau_1 = \tau_1(\eta_1) = 2\eta_1$.

If the rotation $L \to (-L')$ is governed by a junior form (by a senior form), we call the rotation junior (senior).

Lemma V.5.1: *If $L \in G_k$, $-L_1 \in G_k$, $-L_2 \in G_k$ and $L_1 \neq L_2$, then the two rotations $L \to (-L_1)$ and $L \to (-L_2)$ cannot be simultaneously junior.*

Proof: Suppose on the contrary that we have

$$b_1 + L = A_1 C_1,$$

$$b_2 + L = A_2 C_2; \quad C_1 L C_1^{-1} = \bar{A}_1 L \bar{A}_1^{-1} = -L_1;$$

$$C_2 L C_2^{-1} = \bar{A}_2 L \bar{A}_2^{-1} = -L_2;$$

$$\det(A_i) = -a_i < 0; \quad \det(C_i) = -c_i < 0 \qquad (i = 1, 2);$$

$$\varphi_i = (a_i, b_i, c_i) \in m \qquad (i = 1, 2).$$

Setting in (V.4.2) and (V.4.4) $\zeta_1 = 0.1\eta_1$, we get

$$L = 0(D^{\frac{1}{4}+1.1\eta_1}); \quad L_i = 0(D^{\frac{1}{4}+1.1\eta_1}).$$

From Lemma V.3.1 (formula (V.3.9)) we obtain

$$A_i = 0(D^{1.1\eta_1}\sqrt{a_i}); \quad C_i = 0(D^{1.1\eta_1}\sqrt{c_i}) \qquad (i = 1, 2). \tag{V.5.1}$$

Further, from the equations of the system G_k we deduce in the same manner the approximation

$$P_k = 0(D^{1.1\eta_1} p^{s/2}). \tag{V.5.2}$$

In the equations of the system G_k: $l + L_i = P_k Q_i$ we may assume that $\det(Q_i)$ is not divisible by p. If this is not so, then substituting $P_n \to P_k$; $Q_i \to \bar{P}_n + Q_i$ reduces the equation to this type.

We have now

$$l + L = P_k Q_i;$$

$$l - L_j = \bar{P}_k Q_j \qquad (j = 1, 2);$$

$$C_j (l + L) C_j^{-1} = l - L_j,$$

thus in particular

$$C_1 P_k Q_i = \bar{P}_k Q_j C_1.$$

We determine integral X and Y such that

$$\bar{P}_k X + Q_i Y = 1.$$

This is possible because p does not divide $\det(Q_i)$. We conclude now from the preceding equations

$$C_1 P_k = C_1 P_k(\bar{P}_k X + Q_i Y) = \bar{P}_k(P_k C_1 X + Q_j C_1 Y) = \bar{P}_k C_1'.$$

The same reasoning is valid for $\bar{A}_1, C_2, \bar{A}_2$, so that we obtain four equations

$$\text{(V.5.3)} \qquad \begin{aligned} C_1 P_k &= \bar{P}_k C_1'; & \bar{A}_1 P_k &= \bar{P}_k A_1'; \\ C_2 P_k &= \bar{P}_k C_2'; & \bar{A}_2 P_k &= \bar{P}_k A_2'. \end{aligned}$$

We see, therefore, that C_j and A_j ($j=1, 2$) lie on the left versor ray $(\bmod \bar{P}_k)$. In consideration of the properties of such rays, given in § 2 of Chapter II and at the beginning of this chapter, and in view of $\mathrm{Sp}(XY) = \mathrm{Sp}(YX)$, we have

$$\mathrm{Sp}(P_k C_1) \equiv \mathrm{Sp}(P_k C_2) \equiv \mathrm{Sp}(P_k \bar{A}_1) \equiv \mathrm{Sp}(P_k \bar{A}_2) \equiv 0 \pmod{p^s}.$$

Further on the basis of (V.5.1) and (V.5.2),

$$\text{(V.5.4)} \qquad P_k C_j = 0(D^{2.2\eta_1}(c_j p^s)^{\frac{1}{2}}); \quad P_k \bar{A}_j = 0(D^{2.2\eta_1}(a_j p^s)^{\frac{1}{2}}).$$

Since $\varphi_j = (a_j, b_j, c_j) \in m$, we get

$$c_j \leqq \sqrt{\frac{4}{3} D}; \quad a_j = \frac{D + b_j^2}{c_j} \leqq \frac{7}{3} D^{1-\frac{1}{2}+\tau_1} = 0(D^{\frac{1}{2}+2\eta_1}).$$

Substituting this approximation in (V.5.4), we obtain

$$P_k \bar{A}_j = 0(D^{\frac{1}{2}+4.2\eta_1+\frac{\eta_2}{2}}).$$

In view of (V.3.3) we have

$$\text{(V.5.5)} \qquad P_k A_j = 0(D^{\frac{1}{2}+0.92\eta_2}).$$

Analogously we obtain

$$\text{(V.5.6)} \qquad P_k C_j = 0(D^{\frac{1}{2}+0.72\eta_2}).$$

But from (V.3.4) we see that $p^s > D^{\frac{1}{2}+\eta_2}$, and (V.5.3), (V.5.5), and (V.5.6) lead to the conclusion that for sufficiently large D,

$$\mathrm{Sp}(P_k \bar{A}_j) = \mathrm{Sp}(P_k C_j) = 0 \qquad (j=1, 2).$$

Thus we conclude that

$$\text{(V.5.7)} \qquad \mathrm{Sp}(A_j \bar{P}_k) = 0.$$

Moreover, we have under the conditions of Lemma V.5.1

$$\text{(V.5.8)} \qquad A_j \bar{P}_k C_j = p^s(l_j + L) \qquad (j=1, 2).$$

Set $U_j = A_j \bar{P}_k$; $V_j = P_k C_j$. According to (V.5.6) U_1, V_1, U_2, V_2 are vector matrices.

We prove the following auxiliary theorem.

Let U_1, V_1, U_2, V_2 be vector matrices for which $U_1 V_1$ is not a real number and $U_2 V_2 = \lambda_0 + U_1 V_1$ (in the following $\lambda_0, \lambda_1, \ldots$ will be real numbers). Then U_2 and V_2 are linearly dependent on U_1 and V_1:

$$U_2 = \lambda_1 U_1 + \lambda_2 V_1; \quad V_2 = \lambda_3 U_1 + \lambda_4 V_1.$$

For the proof we note that U_i^2 and V_i^2 are real numbers.

Obviously, $U_2 \neq 0$; replacing V_2 by $V_2 + \mu U_2 = V_2'$ (μ being a real number) and omitting the prime on V_2', we may write the equation above as follows:

$$U_2 V_2 = U_1 V_1 .$$

If $U_2 = \lambda_1 U_1 + \lambda_2 V_1$, then we have

$$V_2 = -\frac{U_2}{\det(U_2)} U_1 V_1 = \lambda_4(\lambda_1 U_1 + \lambda_2 V_1) U_1 V_1 .$$

In view of the fact that $V_1 U_1 V_1 = V_1(\lambda_5 + V_1 U_1) = \lambda_5 V_1 + \lambda_6 U_1$, we find that $V_2 = \lambda_3 U_1 + \lambda_1 V_1$ and the assertion is proved. Therefore let U_2 be linearly independent of U_1 and V_1. Then, because V_2 is a vector matrix, we obtain

$$V_2 = \lambda_7 U_2 + \lambda_8 U_1 + \lambda_9 V_1 ,$$

hence

$$U_2 V_2 = \lambda_{10} + U_2(\lambda_8 U_1 + \lambda_9 V_1) = U_1 V_1$$

and

$$U_2 = (U_1 V_1 - \lambda_{10})(\lambda_8 U_1 + \lambda_9 V_1)^{-1} = \lambda_{11} U_1 + \lambda_{12} V_1 ,$$

since

$$U_1 V_1 U_1 = U_1(\lambda_{13} + U_1 V_1) = \lambda_{14} U_1 + \lambda_{15} V_1 .$$

This contradicts the linear independence of U_2 from U_1 and V_1 and proves our theorem.

Turning to (V.5.7), we conclude from our auxiliary theorem that

$$A_2 \bar{P}_k = \lambda_1 A_1 \bar{P}_k + \lambda_2 P_k C_1$$
$$P_k C_2 = \lambda_3 A_1 \bar{P}_k + \lambda_4 P_k C_1 ,$$

where the λ_i's are real numbers. Taking into account that $A_j \bar{P}_k$ and $P_k C_j$ are vector matrices, so that $\overline{P_k C_j} = -P_k C_j$, and taking conjugates on both sides of the equations, we find that

$$P_k \bar{A}_2 = \lambda_1 P_k \bar{A}_1 - \lambda_2 P_k C_1$$
$$P_k C_2 = -\lambda_3 P_k \bar{A}_1 + \lambda_4 P_k C_1 .$$

Cancelling P_k on the left, we obtain

$$\bar{A}_2 = \lambda A_1 + \mu C_1 ,$$
$$C_2 = \nu A_1 + \varrho C_1 ,$$

where λ, μ, ν, ϱ are real numbers. Further, we take the equations $\bar{A}_1 L = L_1 \bar{A}_1$, $C_1 L = L_1 C_1$, multiply the first one by ν, the second one by ϱ, and add them. We get

$$(\nu \bar{A}_1 + \varrho C_1) L = L_1(\nu \bar{A}_1 + \varrho C_1) ,$$

or $C_2 L = L_1 C_2$, which is not possible because of the condition $C_2 L C_2^{-1} = L_2 \neq L_1$ in Lemma V.5.1. This proves Lemma V.5.1.

In the following steps of the proof of the basic lemma the principal difficulty is contained in the study of the behavior of senior forms.

§ 6. Behavior of Senior Forms

We continue the proof of the basic lemma in § 3. Previously we considered systems of equations of integral matrices:

$$A_k: l + L_i = P_k Q_i \quad (i = a_1 + \cdots + a_{k-1} + 1, \ldots, a_1 + \cdots + a_k),$$
$$\bar{A}_k: l - L_i' = \bar{P}_k Q_i' \quad (i = a_1 + \cdots + a_{k-1} + 1, \ldots, a_1 + \cdots + a_k)$$

and rotations $L_\alpha \to (-L_\beta')$, where L_α was one of the L_i's and L_β' one of the L_j''s. Such pairs $(L_\alpha; (L_\beta'))$ will be called conjugate. The rotations $L_\alpha \to (-L_\beta')$ are governed by negative properly primitive binary forms $\varphi = (-a, b, -c)$; we have already determined the behavior of junior forms $\varphi \in m$, those for which

$$D^{\frac{1}{2} - \tau_1} < c \leqq \sqrt{\frac{4}{3}} D.$$

We shall now take up the behavior of senior forms $\varphi \in M$:

$$1 \leqq c \leqq D^{\frac{1}{2} - \tau_1}.$$

We consider forms which are reduced according to the condition in § 5, namely $a \geqq c \geqq |2b|$. Let $(L_\alpha, (-L_\beta'))$ be a conjugate pair whose rotation is governed by the senior form $(-a, b, -c)$ through the equations $l + L_\alpha = B V_\alpha$; $l - L_\beta' = \bar{B} V_\alpha'$; $\det(B) = p^s$, B being one of the matrices $P_1, P_2, \ldots, P_w$ (see § 4), $b + L_\beta = AC$;

$$C L_\alpha C^{-1} = -L_\beta'; \quad \det(A) = -a; \quad \det(C) = -c;$$
$$|2b| \leqq c \leqq a; \quad 1 \leqq c \leqq D^{\frac{1}{2} - \tau_1}.$$

It is possible that the matrix CB is not primitive; then, according to the remarks at the end of § 2, Chapter II, we find a primitive matrix T such that $C = C'T$, $B = \bar{T}B'$, and $C'B'$ primitive.

Let $\det(T) = p^{s_1}$ (it is always possible to assume $\det(T) > 0$), $0 \leqq s_1 \leqq s$; we shall call the number p^{s_1} the index of imprimitivity.

The following lemma will be necessary.

Lemma V.6.1: *The number of conjugate pairs $(L_\alpha; (-L_\beta'))$ in the distinct systems $(A_k; \bar{A}_k)$ $(k = 1, 2, \ldots, w)$, for which the rotations $L_\alpha \to (-L_\beta')$ are governed by a fixed form $\varphi = (-a, b, -c)$ with*

$$|2b| \leqq c \leqq a; \quad \tfrac{1}{2} D^{\frac{1}{2} - \nu} \leqq c \leqq D^{\frac{1}{2} - \nu}; \quad \tfrac{1}{2} \geqq \nu \geqq \tau_1;$$

is of the order

$$0\left(p^{\frac{s}{2} + s_1} D^{\frac{\nu}{2}} D^{-\frac{1}{4} + 4.5\eta_1}\right) \cdot (D, c)^{\frac{1}{2}}. \tag{V.6.1}$$

We consider the conjugate pair $(L_\alpha; (-L_\beta'))$ and the form (a, b, c) with index of imprimitivity p^{s_1}; $s_1 \geqq 0$. We have

$$b + L_\alpha = AC; \quad C L_\alpha C^{-1} = -L_\beta'.$$

We shall prove that the vector matrix $L_\alpha' = T L_\alpha T^{-1}$ is integral and primitive. We have

$$l + L_\alpha' = l + \bar{T}^{-1} L_\alpha \bar{T} = \bar{T}^{-1} B V_\alpha \bar{T} = B' V_\alpha \bar{T}.$$

Hence it follows that L'_α is integral. It is easy to see that L'_α is primitive and $L'^2_\alpha = -D$. Further,

$$\begin{aligned} & L'_\alpha = T L_\alpha T^{-1} = \bar{T}^{-1} L_\alpha \bar{T};\ L_\alpha = T^{-1} L'_\alpha T; \\ \text{(V.6.2)}\qquad & l + L'_\alpha = B' V_\alpha \bar{T} = B' V'_\alpha;\ C' L'_\alpha C'^{-1} = -L'_\beta; \\ & b + L'_\alpha = A'C';\ \det(A') = -a';\ \det(C') = -c'. \end{aligned}$$

Therefore, in view of $C = C'T$, we conclude that $c = c'p^{s_1}$.

We stipulate further, that T is chosen among its left associates such that the fundamental and admissible vector matrix L'_α fulfills a condition of type (V.4.4):

(V.6.3) $$L'_\alpha = 0(D^{\frac{1}{4} + \eta_1 + \zeta_1});\quad \zeta_1 = 0.1\eta_1.$$

We assume again that from the beginning equations of the system A_k not satisfying (V.6.3) are eliminated; for given $p^{s_1} = \det(T)$, different L_α's produce different L'_α's; the number of them which do not fulfill (V.6.3) is, according to Lemma V.4.1, of the order $0(D^{\frac{1}{2} - \eta_1 - \frac{\zeta_1}{2}})$; the number of distinct indices of imprimitivity p^{s_1} is $0(\ln D)$; therefore our hypothesis is realizable. We assume now that

(V.6.4) $$h_2 \geqq \tfrac{1}{8} D^{\frac{1}{2} - \eta_1}.$$

We continue our examination of the pair $L'_\alpha \to (-L'_\beta)$. Arguing as in § 5, we get $CB = \bar{B}C'$; $\mathrm{Sp}(BC) \equiv 0 \pmod{p^s}$;

$$C = 0(D^{\frac{1}{4} - \frac{\nu}{2} + 1.1\eta_1});\quad B = 0(D^{\frac{1}{4} + \frac{\eta_2}{2} + 1.1\eta_1}),$$

$d^s \geqq D^{\frac{1}{2} + \eta_2} = D^{\frac{1}{2} + 10\eta_1}$; hence, as in § 5, we deduce $\mathrm{Sp}(BC) = 0$.

Since $\mathrm{Sp}(C'B') = \mathrm{Sp}(B'C') = p^{-s_1}\,\mathrm{Sp}(BC)$, we have

(V.6.5) $$\mathrm{Sp}(C'B') = \mathrm{Sp}(B'C') = 0,$$

so that $C'B'$ is a vector matrix with $\det(C'B') = -p^{s-s_1}c' < 0$, $C'B'$ primitive, and $C'B' = -\bar{B}'\bar{C}'$.

We prove now that, in the equation $l + L'_\alpha = B'V'_\alpha$,

(V.6.6) $$\mathrm{Sp}(V'_\alpha \bar{C}') \equiv 0 \pmod{c'}.$$

We have

$$\begin{aligned} \bar{B}'\bar{C}' V'_\alpha \bar{C}' &= -C'(B'V'_\alpha)\bar{C}' = -C'(l + L'_\alpha)\bar{C}' \\ &= -l\det(C') - \det(C')\cdot C' L_\alpha C'^{-1} \\ &= c'(l + L'_\beta) \equiv 0 \pmod{c'}. \end{aligned}$$

Further, $C'B'$ is primitive, so that right g.c.d. (C', B') is equal to unity. Choose integral matrices X and Y such that $X\bar{B}' + YC' = 1$. Using the preceding calculation we obtain

$$\bar{C}' V'_\alpha \bar{C}' = (X\bar{B}' + YC')\,(\bar{C}' V'_\alpha \bar{C}') \equiv 0 \pmod{c'}.$$

Hence

$$V'_\alpha \bar{C}' = C' \bar{V}''_\alpha \quad\text{and}\quad \mathrm{Sp}(V'_\alpha \bar{C}') \equiv 0 \pmod{c'},$$

which was to be proved.

Therefore, we have $C'B'=H'$ (a vector matrix), and

$$V'_\alpha \overline{C}' = H_2 = c'd + H,$$

where d is an integer and H a vector matrix,

$$\det(H_1) = -c'p^{s-s_1} = -h_1 < 0, \tag{V.6.7}$$

$$\det(H_2) = -v'c' = -h_2 < 0. \tag{V.6.8}$$

Now we pass over to the algebra $\mathscr{E}$. Our integral matrices go over into integral hermitions (whose components are either integers or halves of odd integers, as described in § 2 of Chapter II). We get $C'B' = H_1 = g'_1 i_1 + g'_2 i_2 + g'_3 i_3$; H_1 primitive,

$$V'_\alpha \overline{C}' = H_2 = \frac{dc'}{2} + g''_1 i_1 + g''_2 i_2 + g''_3 i_3;$$

d integral.

Consider the quadratic form

$$\mathrm{Norm}(\overline{H}_1 x + H_2 y) = (\overline{H}_1 x + H_2 y)(H_1 x + \overline{H}_2 y).$$

On the one hand it is equal to

$$f(x, y) = -h_1 x^2 + 2c'lxy - h_2 y^2.$$

On the other hand, this is the same form as

$$\left(\frac{dc'}{2}\right)^2 y^2 + (g'_1 x + g''_1 y)^2 - (g'_2 x + g''_2 y)^2 - (g'_3 x + g''_3 y)^2,$$

or

$$\begin{aligned} &-4h_1 x^2 + 2(4c'l)xy - (4h_2 + d^2 c'^2) y^2 \\ &= (2g'_1 x + 2g''_1 y)^2 - (2g'_2 x + 2g''_2 y)^2 - (2g'_3 x + 2g''_3 y)^2. \end{aligned} \tag{V.6.9}$$

Therefore we have a representation of a negative binary form $f_1(x, y)$ by an indefinite ternary form $X^2 - Y^2 - Z^2$. Here we note that because H_1 is primitive, $(2g'_1, 2g'_2, 2g'_3)$ is equal to 1 or 2. In view of $(l, p) = 1$, we have

$$(4h_1, 4c'l) = 4c'; \quad 4h_2 + d^2 c'^2 \equiv 0 \pmod{c'};$$

thus a divisor δ of the binary form $f_1(x, y)$ will be equal to c' or $4c'$.

We now investigate the number of original conjugate pairs $(L_\alpha; (-L'_\beta))$ corresponding to given $d, g'_1, g'_2, g'_3, g''_1, g''_2, g''_3$, taking into account that L_α and L'_β are admissible and fundamental. Given these quantities, H_1 and H_2 are defined. Further, $H_1 = C'B'$; H_1 is primitive. If C'_0, B'_0 determine a solution of the equation $C'B' = H_1$, then all other solutions will have the form $C'_0\varepsilon$, $\bar{\varepsilon}B'_0$. Further, $V'_\alpha \overline{C}' = H_2$, so that $V'_\alpha \bar{\varepsilon} \overline{C}'_0 = H_2$; $V'_\alpha = V'_{\alpha_0}\varepsilon$. Moreover, it is necessary that $B'V'_\alpha = l + L'_\alpha$, where L' is a fundamental admissible hermition. Hence $\bar{\varepsilon}B'_0 V'_{\alpha_0}\varepsilon = l + L'_\alpha$; L'_α being fundamental and admissible.

According to what we proved in § 2, this defines $\pm\varepsilon$, if $\mathrm{Norm}(\varepsilon) = +1$. But $\mathrm{Norm}(C') = \mathrm{Norm}(C'_0\varepsilon) = -c'$, which results in $\mathrm{Norm}(\varepsilon) = +1$.

Thus C' may take on no more than two values and determines all remaining hermitions. Also, $L'_\beta = -C'L'_\alpha C'^{-1}$ is determined.

Finally, $L_\alpha = T^{-1}L'_\alpha$ is necessarily fundamental and admissible and $\det(T) = p^{s_1}$. This defines T up to its sign and gives us two possible values for L_α. Therefore, to the given numbers $d, g'_1, \ldots, g''_3$ there correspond no more than four conjugate pairs $(L_\alpha; (-L'_\beta))$.

We return to the representation (V.6.9) and prove that c' divides the greatest common divisor of the three determinants

(V.6.10) $$\begin{vmatrix} 2g'_2 & 2g''_2 \\ 2g'_3 & 2g''_3 \end{vmatrix}, \quad \begin{vmatrix} 2g'_3 & 2g''_3 \\ 2g'_1 & 2g''_1 \end{vmatrix}, \quad \begin{vmatrix} 2g'_1 & 2g''_1 \\ 2g'_2 & 2g''_2 \end{vmatrix}.$$

We have

$$\begin{aligned} 4c'V'B' &= (2V'_\alpha \bar{C}')(2C'B) = 2H_1 \cdot 2H_2 \\ &= 4\mathcal{R}(H_1H_2) + (2dc'g'_1 + 2g''_2 2g''_3 — 2g''_3 \cdot 2g'_2)\, i_1 + \\ &+ (2dc'g'_2 + 4g''_1 g'_3 - 4g''_3 g'_1) i_2 + (2dc'g'_3 + 4g''_2 g'_1 - 4g'_2 g''_1) i_3 ; \end{aligned}$$

therefore the above statement follows.

Without restrictions on the components there may be infinitely many representations of type (V.6.9). If H_1 and H_2 are vectors with negative norms, then the equation $H_1H_2 = \xi + M$, where M is a vector, ξ an integer, gives a representation of the form

$$\mathrm{Norm}(H_1)x^2 + 2\xi xy + \mathrm{Norm}(H_2)y^2$$

by the form $X^2 - Y^2 - Z^2$.

If $\mathrm{Norm}(\varepsilon) = +1$, then we have $\varepsilon H_1 \bar{\varepsilon} \cdot \varepsilon H_2 \bar{\varepsilon} = \xi + \varepsilon M \bar{\varepsilon}$, which also produces such a representation for any integral ε. Representations as (V.6.9) are typically bounded by restrictions on the components. We shall give such bounds. The number d must assume a definite set of values which we approximate in the following. The magnitude of H_1 is independent of this approximation. We have

$$H_1 = C'B' ; \quad C' = CT^{-1} = \frac{C\bar{T}}{\mathrm{Norm}(T)} .$$

Further,

$$L'_\alpha = TL_\alpha T^{-1} ; \quad L_\alpha = 0(D^{\frac{1}{2}+1.1\eta_1}) ; \quad L'_\alpha = 0(D^{\frac{1}{2}+1.1\eta_1}) .$$

Hence from Lemma V.3.1

$$T = 0(p^{\frac{s_1}{2}} \cdot D^{1.1\eta_1}) .$$

We see from (V.6.2), in view of

$$L_\beta = 0(D^{\frac{1}{2}+1.1\eta_1}), \quad C = 0(c^{\frac{1}{2}} D^{1.1\eta_1}) ,$$

so that

(V.6.11) $$C' = p^{-s_1} 0(p^{\frac{s_1}{2}} c^{\frac{1}{2}} D^{2.2\eta_1}) = 0(c^{\frac{1}{2}} p^{-\frac{s_1}{2}} D^{2.2\eta_1}) .$$

By similar calculations,

(V.6.12) $$B' = 0(p^{\frac{s-s_1}{2}} D^{2.2\eta_1}) ,$$

so that

(V.6.13) $$H_1 = 0(c^{\frac{1}{2}} p^{\frac{s}{2}-s_1} D^{4.4\eta_1}).$$

Therefore, we have the representation (V.6.9) subject to the conditions

(V.6.14) $$2g'_1;\quad 2g'_2;\quad 2g'_3 = 0(c^{\frac{1}{2}} p^{\frac{s}{2}-s_1} D^{4.4\eta_1}),$$

(V.6.15) $$(2g'_1, 2g'_2, 2g'_3) = 2 \text{ or } 1.$$

A divisor δ of the binary form $f_1(x, y)$ divides four times the greatest common divisor of the three determinants in (V.6.10). It will be useful to give an estimate of the number of representations (V.6.9) under these conditions. We shall make use of known facts from the Gaussian theory of quadratic forms (see, for instance, VENKOV [2]).

Let $e > 1$ be the greatest common divisor of the determinants in (V.6.10). It is known that e^2 divides Δ, where Δ is the determinant of the negative form $f_1(x, y)$. The number of possible values of e is then $0(\Delta^{\varsigma})$; we approximate the number of representations corresponding to the fixed quantity e.

The greatest common divisor of the integers $2g'_1, 2g'_2, 2g'_3$, which we denote by χ, is equal to 1 or 2. We set $e = \chi\mu$ and consider the two systems of matrices S_λ

(V.6.16) $$\begin{pmatrix} 1 & -\frac{\lambda}{e} \\ 0 & \frac{1}{e} \end{pmatrix} \qquad (\lambda = 0, 1, 2, \ldots, e-1),$$

(V.6.17) $$\begin{pmatrix} \frac{1}{2} & -\frac{\lambda}{e} \\ 0 & \frac{2}{e} \end{pmatrix} \qquad (\lambda = 0, 1, 2, \ldots, \tfrac{e}{2}-1).$$

The first system corresponds to the values $\chi = 1$, $\mu = e$ and the second to $\chi = 2$, $\mu = \frac{e}{2}$.

We transform the form $f_1(x, y)$ with these matrices and discard all non-integral resulting forms. For the integral forms $f_1 S_\lambda = \Psi_\lambda$ we look for all primitive representations by the form $X^2 - Y^2 - Z^2$; their six components a, a', a'', b, b', b'' will differ in that

(V.6.18) $$a = \frac{2g'_1}{\chi};\quad a' = \frac{2g'_2}{\chi};\quad a'' = \frac{2g'_3}{\chi};\quad \chi = 1 \text{ or } 2.$$

We estimate the number of values λ for which the form Ψ_λ is integral. We examine the cases $\chi = 1$ and $\chi = 2$. Let $\chi = 1$. We set further $4p^{s-s_1} = -\alpha_1$; $4e = \beta_1$; $4v' + d^2c' = \gamma_1$, so that

$$f_1(x, y) = c'(\alpha_1 x^2 + 2\beta_1 xy + \gamma_1 y^2);\quad (\alpha_1, \beta_1, \gamma_1) \leqq 4.$$

Moreover, $4e$ is divisible by c'. We set $4e = e_1 c'$. The form $f_1 S_\lambda = \Psi_\lambda$ is necessarily integral, hence we get the congruences

(V.6.19) $$\begin{cases} 4(\alpha_1\lambda - \beta_1) \equiv 0 \pmod{e_1} \\ 16(\alpha_1\lambda - \beta_1)^2 \equiv -\Delta/c'^2 \pmod{e_1^2 c'}. \end{cases}$$

Hence we conclude easily that the number of possible values of λ is of the order

$$0(e^{\zeta})\,(\Delta/c'^2, c')^{\frac{1}{2}}.$$

But since $\Delta = 16c'^2(l^2 - h_1 h_2)$, we deduce by definition of h_1 and h_2

$$(\Delta/c'^2, c') \leqq 32(D, c').$$

Therefore, the required number of values of λ is of the order $0(e^{\zeta})\,(D, c')^{\frac{1}{2}}$. But $(D, c') = (D, c)$, so that we have the approximation $0(e^{\zeta})\,(D, c)^{\frac{1}{2}}$.

The case $\chi = 2$ is completely analogous to the first one. Therefore the number of integral forms Ψ_λ for given e is of the order $0(e^{\zeta_1})\,(D, c)^{\frac{1}{2}}$.

We estimate now the number of values d. We have $c'd = \mathcal{R}(V'_\alpha C')$. Further, $l + L'_\alpha = B' V'_\alpha$; here $l = 0(p^s)$; $L'_\alpha = 0(D^{\frac{1}{2} + 1.1\eta_1})$. Since $\eta_2 = 10\eta_1$, $p^s > D^{\frac{1}{2} + \eta_2}$, we have $l + L'_\alpha = 0(p^s)$. By (V.6.16)

$$B' = 0(p^{\frac{s - s_1}{2}} D^{2.2\eta_1}),$$

so that

$$V'_\alpha = (l + L'_\alpha)\,\frac{\bar{B}'}{\operatorname{Norm}(B')} = 0(p^{\frac{s + s_1}{2}} D^{2.2\eta_1}). \tag{V.6.20}$$

Making use of (V.6.11), we find that

$$V'_\alpha C' = 0(c^{\frac{1}{2}} p^{\frac{s}{2}} D^{4.4\eta_1}). \tag{V.6.21}$$

In view of the equation $c'd = \mathcal{R}(V'_\alpha C')$, we see that

$$d = 0(c'^{-1} c^{\frac{1}{2}} p^{\frac{s}{2}} D^{4.4\eta_1});$$

since $c' = cp^{-s_1}$, we obtain at last

$$d = 0(p^{\frac{s}{2} + s_1} c^{-\frac{1}{2}} D^{4.4\eta_1}). \tag{V.6.22}$$

Hence we deduce an estimate for the determinant Δ of the binary form $f_1(x, y)$ given in (V.6.9). From (V.6.22) and a trivial calculation concerning h_1 and h_2 (see (V.6.7) and (V.6.8)) we conclude that each coefficient of $f_1(x, y)$ is of the order $0(D^2)$, so that $\Delta = 0(D^4)$. Thus the number of divisors e of Δ is estimated by $0(D^{\zeta_1})$.

In view of this the total number of integral forms Ψ_λ for all values of e, according to the preceding calculations, will be of the order $0(D^{\zeta_1})\,(D, c)^{\frac{1}{2}}$.

§ 7. An Estimate for the Number of Primitive Representations

Now we shall estimate the number of primitive representations of any one of the given forms Ψ_λ by $X^2 - Y^2 - Z^2$. In such representations

$$\begin{gathered} \Psi_\lambda = (ax + bx)^2 - (a'x + b'x)^2 - (a''x + b''x)^2 \\ a i_1 + a' i_2 + a'' i_3 = \frac{2g'_1}{\chi} i_1 + \frac{2g'_2}{\chi} i_2 + \frac{2g'_3}{\chi} i_3; \quad \chi = 1 \text{ or } 2, \end{gathered} \tag{V.7.1}$$

according to (V.6.18). In addition (see Venkov [2], p. 141),

$$b i_1 + b' i_2 + b'' i_3 = \frac{2g''_1 - \lambda\cdot 2g'_1}{\mu} i_1 + \frac{2g''_2 - \lambda\cdot 2g'_2}{\mu} i_2 + \frac{2g''_3 - \lambda\cdot 2g'_3}{\mu} i_3.$$

Thus

(V.7.2) $$ai_1 + a'i_2 + a''i_3 = \frac{2H_1}{\chi}$$

(V.7.3) $$bi_1 + b'i_2 + b''i_3 = \frac{M}{\mu} - \frac{\lambda}{\mu} \cdot 2H_1$$

where $M = 2H_2 - dc'$. The numbers λ and $\mu = e$ or $\mu = e/2$ are taken from (V.6.16) or (V.6.17).

According to § 6 it is necessary that

(V.7.4) $$H_1 H_2 = c'(l + L'_\beta),$$

where L'_β is a properly primitive and fundamental vector, and $L'^2_\beta = -D$.

In order to find the primitive representations (V.7.1) we shall find all systems of Gaussian numbers M, N characterizing the representation (see VENKOV [2], p. 139). If the determinant of Ψ_λ is denoted by Ω_λ, then M and N must satisfy in this case the congruences $-p \equiv N^2$; $q \equiv MN$; $-r \equiv M^2 \pmod{\Omega_\lambda}$, where $\Psi_\lambda = (p, q, r)$. The set of numbers $M, N \bmod \Omega_\lambda$ will be of the order $0(D^{\zeta_1})$ (see VENKOV [2], p. 131). We consider any pair of numbers M, N and the representations (V.7.1) corresponding to $\pm M$, $\pm N$. In the worst case for the estimate above the ternary form g constructed for the numbers M, N (see VENKOV [2], p. 131) will be integral (this will always be true in the given case) and equivalent to the form $f = X^2 - Y^2 - Z^2$. If $fS = g$, where

$$S = \begin{pmatrix} a & b & c \\ a' & b' & c' \\ a'' & b'' & c'' \end{pmatrix}$$

is a unimodular integral matrix, then all primitive representations (V.7.1) belonging to $\pm M$, $\pm N$ will be given by its first two columns. It is known that if $fS = g$; $fS_1 = g$, then SS_1^{-1} is an automorphism of f.

Denoting by U, U', ... automorphisms of f, we obtain

$$SS_1^{-1} = U; \quad S_1 = U^{-1}S = U'S = U' \begin{pmatrix} a & b & c \\ a' & b' & c' \\ a'' & b'' & c'' \end{pmatrix}.$$

The obtained representation Ψ_λ will be equivalent to the representation obtained by letting the automorphism act on $X^2 - Y^2 - Z^2$; setting in it

(V.7.5) $$X = ax + by; \quad Y = a'x + b'y; \quad Z = a''x + b''y.$$

We set $H'_1 = ai_1 + a'i_2 + a''i_3$, $M' = bi_1 + b'i_2 + b''i_3$. Then

$$X^2 - Y^2 - Z^2 = \mathrm{Norm}(Xi_1 + Yi_2 + Zi_3) = \mathrm{Norm}(H'_1 x + M'y);$$

where X, Y, Z are replaced according to (V.7.5). We want to prove that the result of the two operations described above is equivalent to the substitution

$$H'_1 \to \varepsilon H'_1 \varepsilon^{-1}; \quad M' \to \varepsilon M' \varepsilon^{-1}; \quad \mathrm{Norm}(\varepsilon) = \pm 1.$$

The general form of integral automorphisms of the form $X^2 - Y^2 - Z^2$ with determinant $+1$ is given by a formula of type (V.2.14) (see BACHMANN, Ch. I, pp. 5—12; this reference does not deal with hermitions, but the corresponding formula for automorphisms is easily deduced from those given):

(V.7.6) $$X' i_1 + Y' i_2 + Z' i_3 = \varepsilon (X i_1 + Y i_2 + Z i_3) \varepsilon^{-1},$$

where $\mathrm{Norm}(\varepsilon) = \pm 1$; ε an integral hermition.

In view of the considerations above, we obtain from the representation of Ψ_λ by f, corresponding to the substitution S, all other representations corresponding to the product $\varepsilon(H_1' x + M' y)\varepsilon^{-1}$ or

$$\varepsilon H_1' \varepsilon^{-1} \cdot x + \varepsilon M' \varepsilon^{-1} \cdot y .$$

Thus, H_1' is replaced by $\varepsilon H_1' \varepsilon^{-1}$. We turn now to (V.7.2) and (V.7.3). H_1 is replaced by $\varepsilon H_1 \varepsilon^{-1}$ and from (V.7.3),

$$H_2 = \frac{\mu}{2} M' + 2\lambda H_1 + dc',$$

so that H_2 goes over into $\varepsilon H_2 \varepsilon^{-1}$ (since $\varepsilon dc' \varepsilon^{-1} = dc'$).

In view of this the product $H_1 H_2 = c'(l + L_\beta')$ will be replaced by $\varepsilon H_1 \varepsilon^{-1} \varepsilon H_2 \varepsilon^{-1} = \varepsilon H_1 H_2 \varepsilon^{-1} = c'(l + \varepsilon L_\beta' \varepsilon^{-1})$.

The vector $\varepsilon L_\beta' \varepsilon^{-1}$ is to be fundamental together with L_β', so that ε must be equal to ± 1.

From the material presented in § 6 and § 7 we conclude that for given d the number of representations (V.6.9), under the bounds given in this paragraph, will be of the order $0(D^\zeta)\,(D, c)^{\frac{1}{2}}$; $\zeta > 0$ being an arbitrarily small constant.

For given s, d we have the estimate (V.6.22). Therefore, taking $\zeta = 0.1\,\eta_1$, we conclude that for given h_1, h_2, c', l and variable d the number of equations (V.6.9) is of the order

(V.7.7) $$0(p^{\frac{s}{2} + s_1} c^{-\frac{1}{2}} D^{4.5\,\eta_1})\,(D, c)^{\frac{1}{2}} .$$

We observe now that $h_1 = c' p^{s - s_1} c\, p^{s - 2 s_1}$ is determined by s_1 and c; l is fixed for given s; $h_2 = v' c' = v' c p^{-s_1}$. Further, from $l + L_\alpha' = B' V_\alpha'$ we find that $l^2 + D = p^{s - s_1} \det(V_\alpha') = p^{s - s_1} v'$, thus v' is fixed for given s, s_1.

Therefore, for the form $\varphi = (-a, b, -c)$ we have the estimate (V.7.7) for the number of equations (V.6.9) generating it and according to § 6 an estimate of the number of conjugate pairs $(L_\alpha; (-L_\beta'))$ of the type given in the formulation of Lemma V.6.1. In view of the condition $\frac{1}{2} D^{\frac{1}{2} - \nu} < c \leqq D^{\frac{1}{2} - \nu}$ we obtain from (V.7.7) the estimate

(V.7.8) $$0(p^{\frac{s}{2} + s_1} D^{\frac{\nu}{2}} D^{-\frac{1}{4} + 4.5\,\eta_1})\,(D, c)^{\frac{1}{2}},$$

which coincides with (V.6.1).

We return now to the proof of the basic lemma. We have seen that the number of reduced properly primitive forms (c, b, a) (with $2b \leqq c \leqq a$) with given c is of the order $0(D^\zeta \sqrt{(D, c)})$. This will obviously also be an estimate for the number of our negative forms $(-a, b, -c)$ for given c. Further, for given $s_1 c = c' p^{s_1}$ and c' fulfilling the inequalities

$$\tfrac{1}{2} D^{\frac{1}{2}-\nu} p^{-s_1} < c' \leqq D^{\frac{1}{2}-\nu} p^{-s_1}, \tag{V.7.9}$$

we possess an expression for the number of possible values of c. Since $(p, D) = 1$, the number of forms for given c' is of the order $0(D^{\zeta} \sqrt{(D, c')})$. The number of c' under condition (V.7.9) with given $(D, c') = r$ will be $0(D^{\frac{1}{2}-\nu} p^{-s_1} r^{-1})$, and the number of corresponding forms $\varphi = (-a, b, -c)$ will be of the order $0(D^{\zeta} r^{\frac{1}{2}})$ so that the total number of forms for such r is $0(D^{\frac{1}{2}-\nu-\zeta} p^{-s_1} r^{-\frac{1}{2}})$. According to Lemma V.6.1, the number of conjugate pairs $(L_\alpha; (-L'_\beta))$ arising from all such forms will be of the order

$$0(p^{\frac{s}{2}} D^{\frac{1}{4}-\frac{\nu}{2}+4.5\eta_1}).$$

Summing over all $r \mid D$ we get an estimate for given ν, s_1:

$$0(p^{\frac{s}{2}} D^{\frac{1}{4}-\frac{\nu}{2}+4.6\eta_1}). \tag{V.7.10}$$

The number ν and s_1 take on $0(\ln D)$ values, so that for all possible cases we obtain the approximation

$$0(p^{\frac{s}{2}} D^{\frac{1}{4}-\frac{\tau_1}{2}+4.7\eta_1}), \tag{V.7.11}$$

since $\nu \geqq \tau_1$. Further, according to (V.3.4),

$$p^{\frac{s}{2}} = 0(D^{\frac{1}{4}+\frac{1}{2}\eta_2});$$

since $\eta_2 = 10\eta_1$, $\tau_1 = 2\eta_1$, we get from (V.7.11) the estimate

$$0(D^{\frac{1}{2}+9\eta_1}). \tag{V.7.12}$$

This is the estimate for the number of senior rotations $L_\alpha \to (-L'_\beta)$.

For the number of junior rotations we have, according to Lemma (V.5.1), the estimate $0(h(-D))$ or $0(D^{\frac{1}{2}+\zeta})$. In fact, for each $L \in A_k$ there cannot be more than one junior rotation to $(-L')$ in the system A_k; this means that the total number of junior rotations $L \to (-L'_\beta)$ does not exceed the total number of vector matrices in the systems A_k, that is,

$$h(-D) = 0(D^{\frac{1}{2}+\zeta}).$$

Therefore the total number of rotations $L_\alpha \to (-L'_\beta)$ for $L_\alpha \in A_k$, $-L'_\beta \in A_k$ and different k is given by (V.7.12).

Now it is not hard to finish the proof of the basic lemma in § 3.

Let the number if different systems A_k satisfy the inequality (V.4.3):

$$w < D^{\frac{1}{2}-\eta_3}.$$

The number of all rotations $L_\alpha \to (-L'_\beta)$ will be equal to

$$\begin{aligned} a_1^2 + a_2^2 + \cdots + a_w^2 &\geqq \frac{1}{w}(a_1 + a_2 + \cdots + a_w)^2 = \frac{h_2^2}{w} \geqq \\ &\geqq \frac{1}{8w} D^{1-2\eta_1} \geqq \frac{1}{8} D^{\frac{1}{2}+\eta_3-2\eta_1} = \frac{1}{8} D^{\frac{1}{2}+10\eta_1}, \end{aligned} \tag{V.7.13}$$

in view of (V.3.7). This contradicts (V.7.12), and the basic lemma is proved.

§ 8. A Lemma on Divisibility of Matrices in Connection with Probabilities of Large Deviations

We now prove a lemma having independent interest. Let $k \geqq 1$ be a fixed integer and $\Pi^{(1)}, \ldots, \Pi^{(g)}$ the collection of all integral, primitive (not divisible by an integer), right non-associated matrices with determinant p^k. We know from § 2, Chapter II, that there are $g=(p+1)p^{k-1}$ such matrices. Let l_0 be chosen such that

$$l_0^2 + D \equiv 0 \pmod{p^k}, \quad 0 < l_0 < p^k,$$

and let $L_1, L_2, \ldots, L_h$ $(h=h(-D))$ be the collection of all fundamental and admissible vector matrices

$$L = \begin{pmatrix} b & -a \\ c & -b \end{pmatrix}; \quad a > 0, \; b^2 - ac = -D.$$

Lemma V.8.1: *For any matrix Π taken from the collection*

(V.8.1) $$\Pi^{(1)}, \Pi^{(2)}, \ldots, \Pi^{(g)},$$

the number $r(\Pi, D)$ of equations of the form

(V.8.2) $$l_0 + L_\alpha = \Pi V_\alpha$$

satisfies the asymptotic relation

(V.8.3) $$r(\Pi, D) \sim \frac{h(-D)}{p^{k-1}(p+1)}$$

as $D \to \infty$.

The proof will be based on the limit theorem for large deviations given in § 3 of Chapter I (Theorem I.3.2).

Let $\zeta_1, \zeta_2, \ldots$ be small positive constants chosen successively in the following. According to the basic lemma in § 3, we choose $\zeta_1 > 0$, which will be fixed in the following; $\zeta_2 = 10\zeta_1$; select s such that

(V.8.4) $$D^{\frac{1}{2}+\zeta_2} \leqq p^{ks} \leqq p^k D^{\frac{1}{2}+\zeta_2}.$$

Let $l \equiv l_0 \pmod{p^k}$; $l^2 \equiv D \pmod{p^{ks}}$; $0 < l < p^{ks}$.

We form the equations of type (V.3.5):

(V.8.5) $$l + L_\alpha = \Pi_\alpha X_\alpha; \qquad (\alpha = 1, 2, \ldots, h).$$

Here $\det(\Pi_\alpha) = p^{ks}$; the Π_α are right non-associated. We expand Π_α in a product of the form

(V.8.6) $$\Pi_\alpha = R_{\alpha 1} R_{\alpha 2} \ldots R_{\alpha s},$$

where $R_{\alpha 1}$ is selected among all right associates of the matrices $\Pi^{(i)}$ such that the integral vector matrix $R_{\alpha 1}^{-1} L_\alpha R_{\alpha 1}$ is fundamental; $R_{\alpha 2}$ is selected such that $(R_{\alpha 1} R_{\alpha 2})^{-1} L_\alpha (R_{\alpha 1} R_{\alpha 2})$ is fundamental, etc. We have $R_{\alpha j} = \Pi^{(i)} \varepsilon$, ε being a unimodular matrix.

We prove that for given j the vector matrices

$$L'_\alpha = (R_{\alpha 1} R_{\alpha 2} \ldots R_{\alpha j})^{-1} L_\alpha (R_{\alpha 1} \ldots R_{\alpha j})$$

are different for different α. Indeed, if $L'_\alpha = L'_\beta$ for $\alpha \neq \beta$, then we have

$$U_{\alpha j} = R_{\alpha 1} \dots R_{\alpha j}; \quad R_{\alpha, j+1} \dots R_{\alpha s} X_\alpha = V_\alpha;$$
$$l + L_\alpha = U_{\alpha j} V_\alpha; \quad l + L'_\alpha = V_\alpha U_{\alpha j};$$
$$l + L_\beta = U_{\beta j} V_\beta; \quad l + L'_\beta = V_\beta U_{\beta j} = V_\alpha U_{\alpha j}.$$

In view of the primitivity of L'_β we find $U_{\beta j} = \varepsilon U_{\alpha j}$, where ε is a unimodular matrix. Hence $U_{\alpha j}^{-1} L_\alpha U_{\alpha j} = U_{\alpha j}^{-1} \varepsilon^{-1} L_\beta \varepsilon U_{\alpha j}$ and $L_\alpha = \varepsilon^{-1} L_\beta \varepsilon$. Since L_α and L_β are fundamental vector matrices, we conclude that $\varepsilon = \pm 1$ and $L_\alpha = L_\beta$, as required. This proves our assertion.

We consider further the matrix

(V.8.7) $$\|R_{\alpha j}\|; \quad \alpha = 1, 2, \dots, h; \quad j = 1, 2, \dots, s;$$

consisting of the matrices $R_{\alpha j}$. According to that proved above, each column of this matrix must represent a permutation of the first column, consisting of the matrices $R_{\alpha 1}$. We denote by $\Psi(\Pi)$ the number of indices α for which $R_{\alpha 1}$ is right associated with $\Pi: R_{\alpha 1} = \Pi \varepsilon$. This quantity will coincide with the number of $R_{\alpha j}$ right associated with Π for fixed j (chosen from 1, ..., s) and $\alpha = 1, 2, \dots, h$. We divide the matrices (V.8.7) for a given small $\zeta_0 > 0$ into three types.

Type I: $\Psi(\Pi)$ fulfills the inequalities

(V.8.8) $$(1 - \zeta_0) h/g \leqq \Psi(\Pi) \leqq (1 + \zeta_0) h/g;$$

Type II:

(V.8.9) $$\Psi(\Pi) > (1 + \zeta_0) h/g;$$

Type III:

(V.8.10) $$\Psi(\Pi) < (1 - \zeta_0) h/g.$$

If for every given $\zeta_0 > 0$ and $D > D_0(\zeta_0, p^k)$ the matrix is of Type I, then (V.8.3) is proved, that is, also Lemma V.8.1. Therefore we assume that for given Π and certain large values of D the matrix (V.8.7) is of Type II or Type III. This hypothesis will be reduced to a contradiction.

First assume the matrix (V.8.7) to be of Type II. We calculate the total number of $R_{\alpha j}$ which are right associated with Π, for all hs values of α and j. Adding them up by columns, we find that the required number is

$$s\Psi(\Pi) > (1 + \zeta_0) \frac{hs}{g}.$$

Now we shall make the same calculation first for rows, then for columns. Let h' be the number of rows of the matrix $\|R_{\alpha j}\|$ in which the matrices right associated with Π occur more than $(1 + \zeta_0/2) s/g$ times. The number of such occurrences is, of course, no larger than s. Hence according to (V.8.9), we find that

$$h's + (h - h')\left(1 + \frac{\zeta_0}{2}\right)\frac{s}{g} \geqq s\Psi(\Pi) > (1 + \zeta_0)\frac{hs}{g}.$$

Therefore we conclude

(V.8.11) $$h' > \frac{\zeta_0 h}{2g-2-2\zeta_0} > \frac{\zeta_0 h}{2g}.$$

Now let the matrix $\|R_{\alpha j}\|$ be of Type III and let h'' be the number of its rows in which matrices right associated with Π occur less than $(1-\zeta_0/2)\,s/g$ times. In this case we have

$$s\,\Psi(\Pi) < (1-\zeta_0)\,\frac{hs}{g}.$$

In $h-h''$ rows the number of occurrences described above will not be less than $(1-\zeta_0/2)\,s/g$. In view of this, $s\,\Psi(\Pi) \geqq (h-h'')\,(1-\zeta_0/2)\,s/g$, so that

$$(1-\zeta_0)\,\frac{sh}{g} \geqq (h-h'')\left(1-\frac{\zeta_0}{2}\right)\frac{s}{g},$$

and

(V.8.12) $$h'' \geqq \frac{\zeta_0}{2-\zeta_0}\,h > \frac{\zeta_0}{2}\,h.$$

If the matrix $\|R_{\alpha j}\|$ is of Type III, we pick out h'' of its rows as described above; h'' is bounded from below by the inequality (V.8.12). We number these indices $\alpha = 1, 2, \ldots, h''$, and obtain h'' equations of the form

(V.8.13) $$l + L_\alpha = R_{\alpha 1} \ldots R_{\alpha s} V_\alpha.$$

The products $\Pi_\alpha = R_{\alpha 1} \ldots R_{\alpha s}$ will contain less than $(1-\zeta_0/2)\,s/g$ matrices $R_{\alpha j}$ which are right associated with Π.

We need an upper estimate for the total number of primitive matrices of this type. In Chapter IV, § 5, we considered an estimate of this sort for products of quaternions. The results of that paragraph are carried over without difficulty to products of matrices. We obtain the following estimate for the number w' of different primitive products $\Pi_\alpha = R_{\alpha 1} \ldots R_{\alpha s}$:

(V.8.14) $$w' < c(\zeta_0)\cdot W \cdot p^{-ks\cdot\eta(\zeta_0)}$$

where $\eta(\zeta_0) > 0$, $c(\zeta_0)$ and ζ_0 are constants, and W is the total number of primitive matrices Π_α, here equal to $(p+1)\,p^{ks-1}$.

The introduction of matrices instead of quaternions does not change the proof. If the matrix $\|R_{\alpha j}\|$ is of Type II, then we obtain the same estimate of the number of different Π_α.

In view of this, if the matrix $\|R_{\alpha j}\|$ is for example of Type III, then among the $h'' > \frac{\zeta_0}{2} h$ equations (V.8.13) there will be $w \leqq w'$ different products Π_α, for which, according to (V.8.14),

(V.8.15) $$w < c_1(\zeta_0)\,p^{ks[1-\eta(\zeta_0)]}.$$

Turning to (V.8.4) we find that

(V.8.16) $$w < c_2(\zeta_0)\,D^{(\frac{1}{2}+\zeta_2)[1-\eta(\zeta_0)]}.$$

The constants in this estimate are also dependent on the number k, which was fixed at the beginning of § 8.

Now the values of the constants will be determined. First we assume ζ_0 arbitrarily small. To it we find $\zeta_2 = 0.01\eta(\zeta_0)$. For sufficiently large D we obtain from (V.8.16)

(V.8.17) $$w < D^{\frac{1}{2} - \frac{1}{4}\eta(\zeta_0)}.$$

We now apply the basic lemma (our number ks corresponds to the number s in this lemma). Setting $\zeta_1 = 0.1$; $\zeta_2 = 0.001\eta(\zeta_0)$; $\zeta_3 = 12\zeta_1 = 0.012\eta(\zeta_0)$, we see from (V.8.17) that

(V.8.18) $$w < D^{\frac{1}{2} - \zeta_3}.$$

Finally, $h'' > \frac{\zeta_0}{2} h = \frac{\zeta_0}{2} h(-D)$. By the theorem of K. L. Siegel, $h(-D) > c_2(\zeta)\, D^{\frac{1}{2} - \zeta}$. Let $\zeta = \frac{1}{2}\zeta_1$; for sufficiently large D we obtain

(V.8.19) $$h'' > D^{\frac{1}{2} - \zeta_1}.$$

An application of the basic lemma to the inequalities (V.8.18) and (V.8.19) leads to a contradiction for sufficiently large D. In the case where the matrix $\|R_{\alpha j}\|$ is of Type II, we obtain the same contradiction. This proves Lemma V.8.1.

From Lemma V.8.1 we deduce Lemma V.8.2, also of independent interest. Denote by $h_1(-D, p^k)$ the number of properly primitive reduced forms (a, b, c) $(|2b| \leqq a \leqq c)$, whose first coefficients are divisible by $p^k : a = a' p^k$.

Lemma V.8.2: *For p and k fixed, $D \to \infty$, we have*

(V.8.20) $$h_1(-D, p^k) \sim \frac{2h(-D)}{p^{k-1}(p+1)}.$$

For the proof we note that if in the reduced properly primitive form (a, b, c) the first coefficient is divisible by p^k, then this is also valid for the form $(a, -b, c)$ inverse to it. From the relations $a = a' p^k$; $l_0^2 \equiv -D \pmod{p^k}$; $b^2 - a' p^k c = -D$; we deduce $l_0^2 - b^2 = (l_0 - b)(l_0 + b) \equiv 0 \pmod{p^k}$. Both factors cannot be divisible by p, because then l_0 would be divisible by p and hence $p|D$, what is impossible. This means, therefore, either $l_0 - b \equiv 0 \pmod{p^k}$ or $l_0 + b \equiv 0 \pmod{p^k}$. Thus, for one and only one of the pair of forms (a, b, c) and $(a, -b, c)$ with $a = a' p^k$, the congruence $l_0 + b' \equiv 0 \pmod{p^k}$ is fulfilled, where b' denotes the middle coefficient of the form. If the forms (a, b, c) and $(a, -b, c)$ coincide, then $b = 0$; the number of such cases is of the order $0(D^{\zeta})$ (see Venkov [2], p. 116).

In Lemma V.8.1 we now take the matrix Π with the special form $\Pi = \begin{pmatrix} p^k & 0 \\ 0 & 1 \end{pmatrix}$ and consider the equations $l_0 + L_\alpha = \Pi V_\alpha$.

Setting

$$L_\alpha = \begin{pmatrix} b_\alpha & -a_\alpha \\ c_\alpha & -b_\alpha \end{pmatrix}, \quad V_\alpha = \begin{pmatrix} x_\alpha & y_\alpha \\ z_\alpha & t_\alpha \end{pmatrix},$$

we find that

$$\begin{pmatrix} l_0 + b_\alpha & -a_\alpha \\ c_\alpha & l_0 - b_\alpha \end{pmatrix} = \begin{pmatrix} p^k & 0 \\ 0 & 1 \end{pmatrix} \begin{pmatrix} x_\alpha & y_\alpha \\ z_\alpha & t_\alpha \end{pmatrix} = \begin{pmatrix} p^k x_\alpha & p^k y_\alpha \\ z_\alpha & t_\alpha \end{pmatrix}.$$

We see that our equations correspond to the form $(a_\alpha, b_\alpha, c_\alpha)$ for which $(a_\alpha, b_\alpha, c_\alpha) = 1$, $a_\alpha = p^k a'_\alpha$; $l_0 + b_\alpha \equiv 0 \pmod{p^k}$.

Therefore, Lemma V.8.1 and the above calculations are sufficient for the proof of (V.8.20).

§ 9. Reduced Forms with Small First Coefficients

Let $x \geqq 1$ be any integer. We denote by $h(-D, x)$ the number of reduced properly primitive binary forms (a, b, c) with $a \leqq x$. The following lemma is then valid.

Lemma V.9.1: *For fixed* $K \geqq 1$,

$$h\left(-D, \frac{\sqrt{D}}{K}\right) \leqq c_1 \frac{h(-D)}{K} \tag{V.9.1}$$

($c_1, c_2, \ldots$ positive constants for given p).

In later paragraphs, instead of (V.9.1), we will be able to prove an asymptotic equation which gives Theorem V.2.2.

Let (a', b', c') be a reduced properly primitive form with $a' \leqq \frac{\sqrt{D}}{K}$. We select the integer k such that

$$\frac{K}{10p} \leqq p^k < \frac{K}{10}. \tag{V.9.2}$$

If k is equal to zero, then $K \leqq 10p$ and (V.9.1) is true for $c_1 = 10p$. We shall assume $K > 10p$; then $k > 0$. We construct the form (p^k, b_1, c_1) and its Dirichlet composition with the form (a', b', c'). We obtain the form $(p^k a', b_2, c_2)$, which may be assumed to be reduced, in view of $p^k a' < \frac{\sqrt{D}}{10}$, and because b_2 may be made to satisfy $|b_2| \leqq a_2/2$. Therefore, to each form (a', b', c') with $a' \leqq \frac{\sqrt{D}}{K}$, we obtain another reduced form (a, b_2, c_2) with $a = p^k a'$; for different (a', b', c') the resulting forms are also different. Hence we have

$$h\left(-D, \frac{\sqrt{D}}{K}\right) \leqq h_1(-D, p^k) < \frac{4h(-D)}{p^{k-1}(p+1)}$$

for large D according to (V.8.20).

Further, by (V.9.2), $p^k \geqq \frac{K}{10p}$, so that

$$h\left(-D, \frac{\sqrt{D}}{K}\right) < \frac{40ph(-D)}{K}$$

for large D. This proves Lemma V.9.1. From this lemma we can deduce some useful relations. We remark that the number of forms (a, b, c) for which $c \geqq k\sqrt{D}$ does not exceed

(V.9.3) $$\frac{c_2 h(-D)}{K}.$$

In fact, we have $ac - a^2/4 \leqq ac - b^2 = D$; $a < \frac{D}{c - a/4}$. For $K > 10$ we get $a \leqq \frac{2D}{c} \leqq \frac{2\sqrt{D}}{K}$ and (V.9.3) follows from (V.9.1).

Further, for $K \geqq 1$ the number of forms (a, b, c) in which $c/a \geqq K$ does not exceed

(V.9.4) $$\frac{c_3 h(-D)}{\sqrt{K}}.$$

Indeed, it follows from $ac - a^2/4 \leqq D$ that

$$a \leqq \frac{\sqrt{D}}{\sqrt{ca^{-1} - \frac{1}{4}}};$$

for $ca^{-1} > K$ we obtain

$$a \leqq \frac{\sqrt{D}}{\sqrt{K - \frac{1}{4}}},$$

so that (V.9.4) follows from (V.9.1).

In the following the quantities described above are given asymptotic expressions with the help of Lobachevskian geometry.

§ 10. Transition to the Proof of Theorem V.2.1

We turn now gradually to the proof of Theorem V.2.1. Let $K_1 > 1$ be an arbitrarily large constant; we consider the quadrangle $A_0(K_1)$ in the fundamental triangle Δ_0. The Lobachevskian straight line $x_2 - K_1 x_1 = 0$ cuts this quadrangle out of Δ_0. The fundamental and admissible points L_j with images in $A_0(K_1)$ will be distinguished in that $c/a \leqq K_1$. According to (V.9.4), the number of such points is

(V.10.1) $$H_0(A_0(K_1)) = h(-D)\left(1 + \theta \frac{c_3}{\sqrt{K_1}}\right)$$

(in the following θ will represent different numbers having absolute value less than or equal to one).

In the following we shall cover $A_0(K_1)$ with a network of quadrangles which do not overlap. Our quadrangle is bounded by the Lobachevskian lines $x_2 - x_1 = 0$; $x_2 - K_1 x_1 = 0$; $-x_1 + 2x_3 = 0$; $x_1 + 2x_3 = 0$.

Let $t > 1$ be a large integer, fixed in the following considerations; we construct the lines

(V.10.2) $$x_2 - \{(v/t)(K_1 - 1) + 1\}\, x_1 = 0 \quad (v = 0, 1, \ldots, t).$$

These lines do not intersect each other and are not parallel. They divide $A_0(K_1)$ into t strips. Further we form the lines

(V.10.3) $$\frac{v_1}{t} x_1 + 2x_3 = 0 \quad (v_1 = -t, -t+1, \ldots, 0, 1, \ldots, t).$$

These lines are parallel. The collections of lines (V.10.2) and (V.10.3) divide $A_0(K_1)$ into quadrangles. A simple calculation demonstrates that for increasing t the Lobachevskian area of these quadrangles converges uniformly to zero. The number of quadrangles for given t is $2t^2 = t_1$.

Let $\Lambda_1, \Lambda_2, \ldots, \Lambda_{t_1}$ be regions obtained from these quadrangles by elimination of certain sides and vertices so that each point of $A_0(K_1)$ lies on one and only one region Λ_m. Here we arrange this elimination such that equally directed vertices (say the "lower left" vertices) lie in Λ_m for all m. These vertices we denote by 0_m. Note that for $t \to \infty$ the Lobachevskian area $\Lambda(\Lambda_m)$ of the region converges uniformly to zero.

Now let Σ_0 be a convex closed region having a piecewise smooth boundary L_0 and lying in $A_0(K_1)$, $A_0(K_1)$ being considered as a bounded part of the Lobachevskian plane. The boundary L_0 is also bounded, moreover uniformly by Σ_0, and lies in $A_0(K_1)$. We shall not prove these facts here. They are elementary properties of convex figures on the Lobachevskian plane. Let a motion be performed on the Lobachevskian plane which carries the vertex 0_m of the region Λ_m into the interior of Σ_0 or its boundary. In this case the whole region Λ_m will lie inside of an expansion Σ_0' of the surface Σ_0, where Σ_0' may be chosen such that

(V.10.4) $$1 \leqq \frac{\Lambda(\Sigma_0')}{\Lambda(\Sigma_0)} \leqq 1 + \alpha(t)$$

where $\alpha(t) \to 0$ as $t \to \infty$ uniformly in m.

We return now the material of § 2. Let the point 0_m on H_0 be represented by the matrix

(V.10.5) $$M = \begin{pmatrix} x_3 & -x_1 \\ x_2 & -x_3 \end{pmatrix}.$$

We consider unimodular matrices $A = \begin{pmatrix} \alpha & \beta \\ \gamma & \delta \end{pmatrix}$ with the property $A^{-1}MA = \bar{A}M\bar{A}^{-1} \in \Sigma_0$. We saw in § 2 that the invariant measure of the set Ω of such matrices on the invariant group is proportional to $\Lambda(\Sigma_0)$. Now we prove a lemma concerning this set.

Lemma V.10.1: *For fixed K_1 and all matrices A we have the bound*

(V.10.6) $$\max(|\alpha|, |\beta|, |\gamma|, |\delta|) < K_2,$$

where

$$K_2 = \tfrac{4}{3}\sqrt{K_1}\,.$$

This lemma is essentially a variant of Lemma V.3.1 and its proof is a complete analogue of the proof of Lemma V.3.1.

Set

$$\bar{A} = \begin{pmatrix} \delta & -\beta \\ -\gamma & \alpha \end{pmatrix} = \begin{pmatrix} \alpha_0 & \beta_0 \\ \gamma_0 & \delta_0 \end{pmatrix};$$

it is sufficient to prove (V.10.6) for $\alpha_0, \beta_0, \gamma_0, \delta_0$. Let

(V.10.7) $$\bar{A} M \bar{A}^{-1} = M' = \begin{pmatrix} x_3' & -x_1' \\ x_2' & -x_3' \end{pmatrix} \in \Sigma_0\,.$$

Since $\Sigma_0 \subsetneqq \Lambda_0$, the points with matrices M and M' correspond to reduced positive binary forms with determinant (-1) (of course, not necessarily integral).

Further, since $\Sigma_0 \subsetneqq A_0(K_1)$, we have

$$1 \leqq x_2/x_1 \leqq K_1$$

and

$$1 \leqq x_2'/x_1' \leqq K_1\,.$$

Moreover,

$$1 = x_1 x_2 - x_3^2 \geqq x_1(x_2 - x_1/4)\,; \quad x_1 x_2 \geqq 1\,.$$

Since $x_1 \geqq x_2/K_1$, we find

$$x_1 \geqq \frac{1}{\sqrt{K_1}}\,; \quad x_2 \leqq \frac{1}{x_1} + \frac{x_1}{4} \leqq \sqrt{K_1} + \frac{x_2}{4}\,; \quad x_2 \leqq \frac{4}{3}\sqrt{K_1}\,;$$

$$x_1 \leqq x_2 \leqq \frac{4}{3}\sqrt{K_1}\,; \quad |x_3| \leqq \frac{x_1}{2} \leqq \frac{2}{3}\sqrt{K_1}\,.$$

The same inequalities are valid for x_1', x_2', x_3'.

We set

$$\varphi(\zeta,\eta) = x_1\zeta^2 + 2x_3\zeta\eta + x_2\eta^2\,; \quad \varphi'(\zeta,\eta) = x_1'\zeta^2 + 2x_3'\zeta\eta + x_2'\eta^2\,.$$

From (V.10.7) we obtain

$$\varphi'(\zeta,\eta) = \varphi(\zeta,\eta)\begin{pmatrix} \alpha_0 & \gamma_0 \\ \beta_0 & \delta_0 \end{pmatrix}.$$

Further,

$$x_1' = x_1\alpha_0^2 + 2x_3\alpha_0\beta_0 + x_2\beta_0^2$$

and

$$x_1 x_1' = (x_1\alpha_0 + x_3\beta_0)^2 + \beta_0^2 \leqq \tfrac{16}{9} K_1$$

in virtue of our estimate.

Hence,

$$|\beta_0| \leqq \tfrac{4}{3}\sqrt{K_1}\,.$$

Analogously,

$$x_2 x_1' = (x_2\beta_0 + x_3\alpha_0)^2 + \alpha_0^2 \leqq \tfrac{16}{9} K_1 \quad \text{and} \quad |\alpha_0| \leqq \tfrac{4}{3}\sqrt{K_1}\,.$$

Considering in exactly the same way the values of x_2x_2' and x_1x_2', we get the same bound for γ_0 and δ_0, and the lemma is proved.

§ 11. A Lemma on Matrices

Let a region Ω of unimodular matrices $A=\begin{pmatrix}\alpha & \beta\\ \gamma & \delta\end{pmatrix}$ bounded by a fixed number of smooth surfaces and satisfying

(V.11.1) $$\max\{|\alpha|,|\beta|,|\gamma|,|\delta|\}\leqq K_2$$

be given.

Let $N>3$ be an odd integer. We introduce the substitution

(V.11.2) $$y_1=\alpha\sqrt{N};\quad y_4=\delta\sqrt{N};\quad y_2=\beta\sqrt{N};\quad y_3=\gamma\sqrt{N}.$$

Then we have $y_1y_4-y_2y_3=N$.

We shall need an asymptotic formula for the number of integral primitive solutions of the relation (V.11.3) with $A\in\Omega$. This number we denote by $f(\Omega,N)$. It is necessary to prove the following lemma, also of independent interest.

Lemma V.11.1: *Let Ω be a region of unimodular matrices $A=\begin{pmatrix}\alpha & \beta\\ \gamma & \delta\end{pmatrix}$ bounded by a fixed number of smooth surfaces and satisfying the conditions*

(V.11.3) $$\max\{|\alpha|,|\beta|,|\gamma|,|\delta|\}\leqq K_2.$$

We consider integral matrices $Y=\begin{pmatrix}y_1 & y_2\\ y_3 & y_4\end{pmatrix}$ with odd determinant $N>1$ and their corresponding unimodular matrices

$$A=\frac{1}{\sqrt{N}}Y.$$

The number $f(\Omega,N)$ of primitive matrices Y for which $A\in\Omega$ satisfies the asymptotic expression

(V.11.4) $$f(\Omega,N)\sim\Psi(N)\,\mathrm{meas}(\Omega)$$

as $N\to\infty$, where $\mathrm{meas}(\Omega)$ is the invariant measure of Ω defined by formula (V.2.15) and $\Psi(N)$ is defined by:

$$\Psi(N)=\frac{6}{\pi^2}V_0(N),$$

(V.11.5) $$V_0(N)=\sum_{r^2\mid N}\mu(r)\,V_1\left(\frac{N}{r^2}\right),$$

(V.11.6) $$V_1(N)=\sum_{\substack{n\mid N\\ n\leqq(\ln N)^{50}}}\frac{N}{n}$$

and $V_0(N)>\frac{1}{4}V_1(N)$.

The proof of this lemma is provided by two basic theorems of I. M. VINOGRADOV and considerations from the arithmetic of matrices.

For this it will be necessary to prove a few lemmas.

Lemma V.11.2: *Let ζ be a small number, $K \geqq 1$ a constant. For the special region Ω_1:*

$$\max\{|\alpha|, |\beta|, |\gamma|, |\delta|\} \leqq K; \quad |\gamma| \leqq \zeta \tag{V.11.7}$$

we have the estimate

$$\operatorname{meas}(\Omega_1) \leqq \frac{64\sqrt{2}}{3} K^3 \zeta. \tag{V.11.8}$$

For the proof we shall use form (V.2.15) for the invariant measure, setting $u_1 = 1$; $u_2 = 2$. Then,

$$\operatorname{meas}(\Omega_1) \leqq \frac{2}{3} \int_{-K\sqrt{2}}^{K\sqrt{2}} d\alpha \int_{-K\sqrt{2}}^{K\sqrt{2}} d\beta \int_{-K\sqrt{2}}^{K\sqrt{2}} d\delta \int_{-\zeta}^{\zeta} d\gamma = \frac{64\sqrt{2}}{3} K^3 \zeta.$$

We shall further consider matrices $Y = \begin{pmatrix} y_1 & y_2 \\ y_3 & y_4 \end{pmatrix}$ with $\det(Y) = N$ and $(y_3, y_4) = 1$. Such matrices will be called variable. We shall deduce an asymptotic expression for the number of variable matrices for which $A = \frac{1}{\sqrt{N}} Y \in \Omega$.

Lemma V.11.3: *The number of right non-associated variable matrices is equal to N. As a set of representatives we can take the right non-associated matrices*

$$\begin{pmatrix} N & \xi \\ 0 & 1 \end{pmatrix}, \quad \xi = 0, 1, \ldots, N-1. \tag{V.11.9}$$

Proof: Let $Y = \begin{pmatrix} y_1 & y_2 \\ y_3 & y_4 \end{pmatrix}$ be variable. We choose z and t such that $y_3 z + y_4 t = 1$ and construct the unimodular matrix

$$\varepsilon = \begin{pmatrix} y_4 & z \\ -y_3 & t \end{pmatrix}.$$

We have $Y \cdot \varepsilon = \begin{pmatrix} N & \xi \\ 0 & 1 \end{pmatrix}$; $\xi = y_1 z + y_2 t$ being integral. Further, if

$$Y = \begin{pmatrix} N & \xi \\ 0 & 1 \end{pmatrix}; \quad Y' = \begin{pmatrix} N & \xi' \\ 0 & 1 \end{pmatrix}; \quad \begin{array}{l} \xi' - \xi \equiv 0 \pmod{N} \\ \xi' = \xi + Nq,\ q \text{ integral}, \end{array}$$

then

$$Y' = Y \begin{pmatrix} 1 & q \\ 0 & 1 \end{pmatrix}.$$

Finally, if $\xi' \not\equiv \xi \pmod N$, then Y and Y' are not right associated, because otherwise

$$Y^{-1}Y' = \begin{pmatrix} 1 & (\xi'-\xi)N^{-1} \\ 0 & 1 \end{pmatrix}$$

would be an integral matrix, which is impossible. The lemma is proved.

Using this lemma it is easy to calculate among other things the number of primitive right non-associated matrices defined in § 10. Thus, for facts which we shall need concerning variable matrices, it is sufficient to investigate all special matrices (V.11.9) and all unimodular matrices ε, for which $\begin{pmatrix} N & \xi \\ 0 & 1 \end{pmatrix}\varepsilon$ satisfies the given conditions. If $A = \frac{1}{\sqrt{N}} Y \in \Omega$, then we write $Y \in \sqrt{N}\,\Omega$.

Lemma V.11.4: *The number of matrices Y with* $\det(Y) = N$ *on the region* $\sqrt{N}\,\Omega_1$, *where* Ω_1 *is defined by* (V.11.7), *is of the order*

$$0(\zeta N \ln^2 N) \tag{V.11.10}$$

as $\zeta \to 0$, $\zeta > \frac{1}{\sqrt{N}}$.

Proof: $y_1y_4 - y_2y_3 = N$. Let $(y_3, y_4) = P$, $y_2P^{-1} = y_2'$, $y_4P^{-1} = y_4'$; $y_1y_4' - y_2'y_3 = NP^{-1}$. Suppose we have the inequality

$$K\sqrt{N}P^{-1}2^{-r-1} \leqq |y_4'| \leqq K\sqrt{N}P^{-1}2^{-r}. \tag{V.11.11}$$

For given y_4', in view of $(y_2', y_4') = 1$, y_3 is defined for given y_2' modulo y_4', and because of (V.11.7) and (V.11.11) assumes $0(\zeta P 2^r)$ values; y_2' assumes $0(\sqrt{N}P^{-1})$ values, and y_1 is defined by the equation, so that all in all we have $0(\zeta NP^{-1})$ values. The number r in (V.11.11) runs through $0(\ln N)$ values if $|y_4'| \geqq 1$. If $y_4' = 0$, the equation has $0(N^\eta)$ solutions. Therefore, the number of solutions is

$$\sum_{P|N} 0(\zeta NP^{-1}\ln N) = 0(\zeta N \ln^2 N), \tag{V.11.12}$$

which was to be shown. We remark also that the proof is valid for $\zeta = K$; we obtain then $0(N\ln^2 N)$ solutions.

Let us consider the special region $\sqrt{N}\,\Omega_1$ of the form

$$\gamma\sqrt{N} \leqq y_3 \leqq (\gamma + \Delta\gamma)\sqrt{N}, \tag{V.11.13}$$

$$\alpha \leqq \frac{y_1}{y_3} \leqq \alpha + \Delta\alpha, \tag{V.11.14}$$

$$\delta\sqrt{N} \leqq y_4 \leqq (\delta + \Delta\delta)\sqrt{N} \tag{V.11.15}$$

where

$$\gamma \geqq (\ln N)^{-100}, \quad \Delta\gamma, \Delta\alpha, \Delta\delta \geqq (\ln N)^{-10} \tag{V.11.16}$$

and $\Delta\gamma$, $\Delta\alpha$, $\Delta\delta$ are small numbers. We shall assume that this region $\sqrt{N}\,\Omega_1$ is contained in $\sqrt{N}\,\Omega$, so that $\max(|\alpha|, |\beta|, |\gamma|) \leqq K$. Let

$$\varepsilon = \begin{pmatrix} x & y \\ z & t \end{pmatrix}$$

be an integral unimodular matrix. We look at the product

$$\text{(V.11.17)} \qquad \begin{pmatrix} N & \xi \\ 0 & 1 \end{pmatrix} \begin{pmatrix} x & y \\ z & t \end{pmatrix} = \begin{pmatrix} Nx+\xi z & Ny+\xi t \\ z & t \end{pmatrix}.$$

The relation $Z \in \sqrt{N}\,\Omega_1$ is equivalent to the inequalities

$$\text{(V.11.18)} \qquad \gamma\sqrt{N} \leqq z \leqq (\gamma + \Delta\gamma)\sqrt{N},$$

$$\text{(V.11.19)} \qquad \delta\sqrt{N} \leqq t \leqq (\delta + \Delta\delta)\sqrt{N},$$

$$\text{(V.11.20)} \qquad \alpha \leqq \frac{Nx}{z} + \xi \leqq \alpha + \Delta\alpha$$

with the conditions

$$\text{(V.11.21)} \qquad (z, t) = 1\,; \quad xt - yz = 1\,.$$

Let us agree to subject x, defined for given z and t by (V.11.21), to the condition $0 < x < z$. Then it is possible to assert the following. For given values of z and t satisfying the condition (V.11.21) the matrix Z is determined.

In fact, for given z and t, x and y can be determined and also the number $\frac{Nx}{z}$. The number ζ is then determined in a unique manner by (V.11.20), and also the matrix Z. On the other hand, we see from Lemma V.11.2 that each variable matrix may be given the form (V.11.17).

From (V.11.20) $\left[\frac{Nx}{z} + \xi\right] = \left[\frac{Nx}{z}\right] + \xi$ is determined. We write $\{\alpha\} = \theta_\alpha$ and allow no integers between α and $\alpha + \Delta\alpha$. Then the inequality

$$\text{(V.11.22)} \qquad \theta_\alpha \leqq \left\{\frac{Nx}{z}\right\} \leqq \theta_\alpha + \Delta\alpha$$

is equivalent to (V.11.20) and Z is uniquely determined. If there are integers in the segment $[\alpha, \alpha + \Delta\alpha]$, then the problem is reduced by an elementary method to one or two inequalities of type (V.11.22); for simplicity we shall stop with the inequalities (V.11.22).

§ 12. A Lemma due to I. M. Vinogradov and Kloosterman Sums

For given z we shall sort out the t for which the conditions (V.11.19) and (V.11.21) are valid, and then we shall find $x \equiv t'(\bmod z)$ from the congruence $xt \equiv 1 (\bmod z)$.

Let T be the set of integers t relatively prime to z and satisfying (V.11.19). We denote by $Q(T, Z)$ the number of $t \in T$ which fulfill (V.11.22) for given Z. The number $Q(T, Z)$ will be calculated by a known method due to I. M. VINOGRADOV. We apply the following lemma (see I. M. VINOGRADOV [1], p. 260, Lemma 12; we substitute Δ' for Δ).

VINOGRADOV'S **Lemma:** *Let r be integral, $r>0$; α, β real, $0<\Delta'<0.05$. Then there exists a periodic function $\Psi(x)$ with period one satisfying the conditions*

1) *$\Psi(x)=1$ in the interval $\alpha+0.5\Delta' \leqq x \leqq \beta-0.5\Delta'$;*
2) *$0 \leqq \Psi(x) \leqq 1$ in the intervals $\alpha-0.5\Delta' \leqq x \leqq \alpha+0.5\Delta'$ and $\beta-0.5\Delta' \leqq x \leqq \beta+0.5\Delta'$;*
3) *$\Psi(x)=0$ in the interval $\beta+0.5\Delta' \leqq x \leqq 1+\alpha-0.5\Delta'$;*
4) *$\Psi(x)$ may be expanded in a Fourier series of the form*

$$\Psi(x)=\beta-\alpha+\sum_{m=1}^{\infty}(a_m\cos 2\pi m x+b_m \sin 2\pi m x),$$

where

$$|a_m|<\frac{2}{\pi m}; \quad |b_m| \leqq \frac{2}{\pi m}; \quad |a_m| \leqq 2(\beta-\alpha);$$

$$|b_m| \leqq 2(\beta-\alpha); \quad |a_m|<\frac{2}{\pi m}\left(\frac{r}{\pi m \Delta'}\right)^r; \quad |b_m|<\frac{2}{\pi m}\left(\frac{r}{\pi m \Delta'}\right)^r.$$

Here we shall take $\alpha=\theta_\alpha$; $\beta=\theta_\alpha+\Delta\alpha$; $\Delta'<0.01\alpha$ is a fixed small number, to be determined in the following; $r=100$. We construct the sum

$$\sum_{t\in T}\bar{\Psi}\left(\frac{Nt'}{z}\right),$$

where $\bar{\Psi}$ denotes the function obtained by applying VINOGRADOV'S lemma for the given values of α and β. We replace α and β by $\theta_\alpha+\Delta'$ and $\theta_\alpha+\Delta\alpha-\Delta'$ respectively and apply the lemma again to obtain a function denoted by $\underline{\Psi}(x)$. We have obviously

$$\text{(V.12.1)} \qquad \sum_{t\in T}\underline{\Psi}\left(\frac{Nt'}{z}\right) \leqq Q(T, z) \leqq \sum_{t\in T}\bar{\Psi}\left(\frac{Nt'}{z}\right).$$

We shall find asymptotic expressions for the left and right sides of (V.12.1). We carry this out for $\bar{\Psi}(x)$; the other expression is obtained analogously. We have

$$\sum_{t\in T}\bar{\Psi}\left(\frac{Nt'}{z}\right)=\Delta\alpha\sum_{t\in T}1+\frac{1}{2}\sum_{m=1}^{\infty}(a_m-ib_m)\sum_{t\in T}\exp\left(\frac{2\pi i N t' m}{z}\right)+$$
$$+\frac{1}{2}\sum_{m=1}^{\infty}(a_m+ib_m)\sum_{t\in T}\exp\left(-\frac{2\pi i N t' m}{z}\right).$$

We denote by $R(N, z)$ the sum of the terms of our series for which $m>N^{0.01}$. Since $r=100$ and in view of the approximations of the coefficients we have

$$\text{(V.12.2)} \qquad R(N, z)=(\Delta')^{-100}\,0(N^{0.005}).$$

The order is uniform in z satisfying (V.11.18). We now set

$$S(m, N, z) = \sum_{t \in T} \exp\left(2\pi i \frac{N t' m}{z}\right);$$

$$M = \sup_{1 \leqq m \leqq N^{0.01}} |S(m, N, z)|.$$

We obtain

$$\text{(V.12.3)} \quad \sum_{t \in T} \bar{\Psi}\left(\frac{N t'}{z}\right) = \Delta\alpha \cdot \sum_{t \in T} 1 + 4\theta(\Delta')^{-100} 0(N^{0.4}) + $$
$$+ 2\theta \cdot \Delta\alpha \cdot N^{0.01} M,$$

where $|\theta| \leqq 1$; θ not always the same.

In the sum

$$S(m, N, z) = \sum_{t \in T} \exp\left(2\pi i \frac{N t' m}{z}\right)$$

the number t runs through an incomplete system of reduced residues (mod z).

We apply the known formula of I. M. VINOGRADOV's for the reduction of sums from an incomplete system of residues to a complete one (see I. M. VINOGRADOV [2]).

This formula is

$$\text{(V.12.4)} \quad \sum_{t \in T} \sum_{\substack{x=1 \\ (x,z)=1}}^{z-1} \sum_{s=0}^{z-1} \exp \frac{2\pi i}{z}(f(x) - (x-t)s) = z \sum_{t \in T} \exp \frac{2\pi i}{z} f(t).$$

Here $f(t)$ is an integral function. To apply this formula we set

$$F_s = \sum_{(x)} \exp \frac{2\pi i}{z}(f(x) - sx)$$

and change the order of summation. We obtain

$$\text{(V.12.5)} \qquad \sum_{t \in T} \exp \frac{2\pi i}{z} f(t) = \frac{1}{z} \sum_{s=0}^{z-1} \sum_{t \in T} F_s \exp \frac{2\pi i t s}{z}.$$

Here we set $f(x) = \dfrac{N x' m}{z}$. Then we have

$$F_s = \sum_{\substack{x=1 \\ (x,z)=1}}^{z-1} \exp \frac{2\pi i}{z}(N m x' - s x).$$

Such a sum is called a Kloosterman sum; estimates (even for more general cases) are found in KLOOSTERMAN, Lemma 4. A better estimate has been given by WEIL. The estimate according to KLOOSTERMAN has the form

$$\min\{0(z^{\frac{3}{4}+\eta}) \cdot (Nm, z)^{\frac{1}{4}};\ 0(z^{\frac{3}{4}+\eta}) \cdot (s, z)^{\frac{1}{4}}\},$$

where η is an arbitrarily small number.

We assume now that

(V.12.6) $$(N, z) \leqq N^{0.01}.$$

In view of $m \leqq N^{0.01}$, we get $(Nm, z) \leqq N^{0.02}$ and taking $\eta = 0.05$,

(V.12.7) $$|F_s| \leqq z^{0.75} N^{0.01}.$$

In (V.12.5) we sum over $t \in T$ for fixed s. Here $(t, z) = 1$ and t satisfies (V.11.19).

Consider the estimate

$$\sum_{t \in T} \exp \frac{2\pi i t s}{z} = \sum (s, T, z).$$

We have (see I. M. Vinogradov [1], p. 254)

(V.12.8) $$\sum (0, T, z) = 0(\sqrt{N}),$$

(V.12.9) $$\sum (s, T, z) = \theta \sum_{r|z} \min \left(\frac{2\delta \sqrt{N}}{r}, \frac{2}{\left\{ \frac{sr}{z} \right\}} \right)$$

for $s \neq 0$. Further, in view of (V.12.7),

$$|F_s| \leqq z^{0.75} N^{0.01},$$

so that (V.12.5) is estimated by

$$0(1) z^{-0.25} N^{0.01} \sum_{r|z} \sum_{s=1}^{z-1} \min \left(\frac{2\delta \sqrt{N}}{r}, \frac{2}{\left\{ \frac{sr}{z} \right\}} \right) +$$
$$+ 0(\sqrt{N}) z^{-0.25} N^{0.01}.$$

The second sum in the first term is of the order

$$0(N^{\eta} \cdot N^{\frac{1}{2}} \ln z),$$

and we get

$$\sum_{t \in T} \exp \frac{2\pi i}{z} f(t) = 0(N^{0.52} z^{-0.25}).$$

On the basis of (V.11.18) and (V.11.16) we have

$$z > N^{\frac{1}{2}} (\ln N)^{-100};$$

so that we finally obtain

(V.12.10) $$\sum_{t \in T} \exp \frac{2\pi i}{z} f(t) = 0(N^{0.4}).$$

The sum on the left side is equal to $S(m, N, z)$, so that $M = 0(N^{0.4})$. Returning to (V.12.3) we find that

$$\sum_{t \in T} \overline{\Psi} \left(\frac{Nt'}{z} \right) = \Delta \alpha \sum_{t \in T} 1 + (\Delta')^{-100} 0(N^{0.4}) + \Delta \alpha 0(N^{0.41}).$$

Setting $\Delta' = (\ln N)^{-1000}$, we get

(V.12.11) $$\sum_{t \in T} \overline{\Psi} \left(\frac{Nt'}{z} \right) = \Delta \alpha \sum_{t \in T} 1 + \Delta \alpha \cdot 0(N^{0.41}).$$

Substituting $\underline{\Psi}$ for $\bar{\Psi}$ and $\Delta\alpha - 2\Delta'$ for $\Delta\alpha$, we obtain the same formula. Hence, because of (V.12.1), we obtain at last

$$\text{(V.12.12)} \qquad Q(T,z) = \Delta\alpha Q(T)\left[1 + 0\left(\frac{1}{(\ln N)^{1000}}\right)\right],$$

where $Q(T) = \sum_{t\in T} 1$. In view of the lower bounds for $\Delta\alpha, \Delta\gamma, \Delta\delta$ in (V.11.16), we get from this a lower estimate.

In order to deduce (V.12.12) we applied the essential bound (V.12.6). Suppose that this is not fulfilled, and thus $(N,z) > N^{0.01}$. We prove then that $Q(T,z)$ is comparatively small. Indeed, let $(N,z) = P > N^{0.01}$. Substituting y_3 for z, y_4 for t and taking into account $(y_3, y_4) = 1$, we get from $y_1 y_4 - y_2 y_3 = N$

$$y_1 \equiv 0 (\operatorname{mod} P)\,; \quad y_1' y_4 - y_2 y_3' = \frac{N}{P}\,;$$

$$y_1' = \frac{y_1}{P}, \quad y_3' = \frac{y_3}{P}\,.$$

Using $P > N^{0.01}$ we conclude from Lemma V.11.4 that the number of matrices with a similar representation in the region $\sqrt{N}\,\Omega_1$ will be $0(N^{0.99}\ln^2 N)$. Further,

$$\sum_{(N,z) > N^{0.01}} Q(T,z) = 0(N^{0.99+\eta})\,.$$

In view of this we may assume that the number of variable matrices in the region $\sqrt{N}\,\Omega_1$ is equal to

$$\sum_{(z)} Q(T,z) + 0(N^{0.99+\eta})\,,$$

taking advantage of the formula (V.12.12) for all z in (V.11.18).

We have further

$$Q(T) = \sum_{t\in T} 1 = \sum_{(t,z)=1} 1$$

for t fulfilling (V.11.19).

Hence

$$Q(T) = \Delta\delta \cdot \sqrt{N} \cdot \frac{\varphi(z)}{z} (1 + 0(N^{-\frac{1}{4}}))\,,$$

where $\varphi(z)$ is Euler's function. Summing over those z described in (V.11.18), we find that

$$\sum_{(z)} Q(T,z) = \Delta\delta \cdot \sqrt{N}\left(\sum_{\gamma\sqrt{N} \leqq z \leqq (\gamma + \Delta\gamma)\sqrt{N}} \frac{\varphi(z)}{z}\right)(1 + 0(N^{-\frac{1}{4}}))\,.$$

Further, we have for $s > 1$

$$\sum_{z=1}^{\infty} \varphi(z) z^{-s-1} = \frac{\varphi(s)}{\varphi(s+1)}\,.$$

For the summation over $\frac{\varphi(z)}{z}$ we integrate the kernel $\frac{x^s}{s}$ along the line $(2-i\infty, 2+i\infty)$. At the pole $s=1$ we have the residue $\frac{x}{\zeta(z)}=\frac{6x}{\pi^2}$. Hence it follows easily that

$$\sum_{\gamma\sqrt{N}\leqq z\leqq(\gamma+\Delta\gamma)\sqrt{N}} \frac{\varphi(z)}{z}=\frac{6}{\pi^2}\Delta\gamma\sqrt{N}+0(N^{0.4}).$$

Substituting the last equation in (V.12.12) we obtain

$$\text{(V.12.13)}\quad \sum_{(z)} Q(T,z)=\frac{6}{\pi^2}\Delta\alpha\,\Delta\delta\,\Delta\gamma\,N\left(1+0\left(\frac{1}{\ln^{100}N}\right)\right),$$

using (V.11.16).

This is a useful expression for the number of variable matrices $Z\in\sqrt{N}\,\Omega_1$. We shall make a few more remarks.

First of all, instead of the condition $\gamma\geqq(\ln N)^{-100}$ we could have taken the condition $\gamma+\Delta\gamma<-(\ln N)^{-100}$. Further, if we replace the less than or equal to or greater than or equal to signs by strictly less than or strictly greater than signs, then we do not change (V.12.13).

We have taken into account only variable matrices in (V.12.13); the condition $(z,t)=1$, that is, $(y_3,y_4)=1$, was imposed. Now let $(y_3,y_4)=P>1$ and $\gamma\geqq(\ln N)^{-50}$, $P\leqq(\ln N)^{50}$. Consider instead of $Y=\begin{pmatrix} y_1 & y_2\\ y_3 & y_4\end{pmatrix}$ the matrix

$$Y'=\begin{pmatrix} y_1 & y_2\\ y_3' & y_4'\end{pmatrix};\quad y_i'=y_iP^{-1}\qquad (i=3,4)$$

and instead of N the number N/P. We observe then that (V.12.13) is valid when we replace N by N/P. If $P\leqq(\ln N)^{50}$, then, as in Lemma V.11.4, the number of corresponding matrices will be $0(N(\ln N)^{-48})$. In view of this, if we denote by $f_1(\Omega_1,N)$ the total number of integral matrices $Z\in\sqrt{N}\,\Omega_1$, then we find, using (V.11.16),

$$f_1(\Omega_1,N)=\frac{6}{\pi^2}\Delta\alpha\,\Delta\delta\,\Delta\gamma\,V_1(N)\left(1+0\left(\frac{1}{\ln^{10}(N)}\right)\right),$$

where

$$V_1(N)=\sum_{\substack{n\mid N\\ n\leqq(\ln N)^{50}}}\frac{N}{n}.$$

It is necessary for us to calculate the quantity $f(\Omega_1,N)$ of primitive matrices $Z\in\sqrt{N}\,\Omega_1$. We consider a non-primitive matrix $Z_1=qZ$, where Z is primitive and q is an integer.

We have then $q^2\mid N$. If $qZ\in\sqrt{N}\,\Omega_1$, then $Z\in\sqrt{\frac{N}{q^2}}\,\Omega_1$ and Z is primitive.

The preceding calculations are valid if we replace N by N/q^2 and if N/q^2 is sufficiently large.

Let $q \leqq N^{\frac{1}{4}}$ so that $N/q^2 \geqq \sqrt{N}$. In this case we can make the above replacement of N/q^2 for N.

If $q > N^{\frac{1}{4}}$, then, according to the remarks in Lemma V.11.4, qZ is of the order $0\left(\frac{N}{q^2}\ln^2 N\right)$.

Therefore we get for primitive matrices

$$\text{(V.12.14)} \quad f(\Omega_1, N) = \frac{6}{\pi^2}\Delta\alpha\,\Delta\delta\,\Delta\gamma\cdot V_0(N)\left(1+0\left(\frac{1}{\ln^{10} N}\right)\right),$$

where

$$V_0(N) = \sum_{r^2|N} \mu(r)\, V_1\left(\frac{N}{r^2}\right).$$

For the proof of this it suffices to show that

$$V_0(N) > \tfrac{1}{4} V_1(N).$$

This follows easily because N is odd and because

$$1 - \sum_{r=3}^{\infty} \frac{1}{r^2} > \frac{1}{4}.$$

The asymptotic expression (V.12.14) proves Lemma V.11.1 for regions Ω of the special type Ω_1. Now we consider the region Ω. We remark that the formula (V.12.14) was proved under the assumption $\gamma \geqq (\ln N)^{-50}$ or $(\gamma + \Delta\gamma) < -(\ln N)^{-50}$. But the numbers γ and δ, z and t, y_3 and y_4 are interchangeable in the previous discussion, so that (V.12.14) remains true if we drop the restriction on γ and require instead that $\delta \geqq (\ln N)^{-50}$ or $(\delta + \Delta\delta) < -(\ln N)^{-50}$.

We return to the surface Ω. Let $\zeta > 0$ be a small constant and $\Omega'(\zeta)$ that part of the region Ω for which $|\gamma| \leqq \zeta$.

In the region $\Omega'(\zeta)$, for ζ sufficiently small, it is not possible that $|\delta| \leqq \zeta$. Otherwise because of the equation $\alpha\delta - \beta\gamma = 1$ we would have $\max(|\alpha|, |\beta|) > K_2$, which is not possible.

Setting $\Omega = \Omega'(\zeta) + \Omega''(\zeta)$, we see that the formula (V.12.14) may be used always for $\Omega_1 \subsetneqq \Omega$. Consider first the region $\Omega''(\zeta)$; the region $\Omega'(\zeta)$ will be treated analogously. The region $\Omega''(\zeta)$ of unimodular matrices satisfies the conditions (V.11.1) and is bounded by a piecewise smooth surface. Inside this surface and on its boundary, $|\gamma| \geqq \zeta$.

In the region $\Omega''(\zeta)$ we employ the parameters $x_1 = \alpha$, $x_2 = \gamma$, $x_3 = \delta$; the parameter β is defined because of $|\gamma| \geqq \zeta$. We introduce these new coordinates, in which the surface Ω of (V.11.15)–(V.11.18) is a straight parallelepiped

$$\alpha' = x_1/x_2\,; \quad \gamma' = x_2\,; \quad \delta' = x_3\,.$$

We cover $\Omega''(\zeta)$ with the grid $\alpha'=\text{constant}$, $\gamma'=\text{constant}$, $\delta'=\text{constant}$. For given small $\zeta_0>0$, if $\Delta\alpha'=\Delta\gamma'=\Delta\delta'=\zeta_1(\zeta_0)$, we get

$$\left|\sum \Delta\alpha'\,\Delta\gamma'\,\Delta\delta' - \iiint_{\Omega''(\zeta)} d\alpha'\,d\gamma'\,d\delta'\right| < \zeta_0 .$$

Here the sum is taken over those parallelepipeds contained in $\Omega''(\zeta)$. Further we have by computation of the Jacobian

$$\frac{D(\alpha',\beta',\gamma')}{D(x_1,x_2,x_3)} = \frac{1}{x_2},$$

hence

$$\iiint_{\Omega''(\zeta)} d\alpha'\,d\beta'\,d\gamma' = \iiint_{\Omega''(\zeta)} \frac{dx_1\,dx_2\,dx_3}{|x_2|}.$$

The interval converges, since $|x_2|\geqq\zeta$. Further, the differential under the integral sign is that of the invariant measure on the unimodular group. Hence, the integral is equal to $\operatorname{meas}(\Omega''(\zeta))$. Turning to formula (V.12.14), we see that for $N\to\infty$ the number of primitive matrices Y with $A=\frac{1}{\sqrt{N}}Y\in\Omega''(\zeta)$ will be

$$f(\Omega''(\zeta),N)\sim\Psi(N)\cdot\operatorname{meas}(\Omega''(\zeta)).$$

The same reasoning for δ in place of γ, and for $\Omega'(\zeta)$ is valid, so that

$$f(\Omega,N)\sim\Psi(N)\operatorname{meas}\Omega$$

and Lemma V.11.1 is proved.

§ 13. Consequences of Lemma V.11.1

Lemma V.11.1 consisted of the following. Let a region Ω of unimodular matrices $A=\begin{pmatrix}\alpha & \beta\\ \gamma & \delta\end{pmatrix}$ bounded by a fixed number of smooth surfaces and satisfying

(V.13.1) $$\max(|\alpha|,|\beta|,|\gamma|,|\delta|)\leqq K_2$$

be given. Consider the integral matrices $Y=\begin{pmatrix}y_1 & y_2\\ y_3 & y_4\end{pmatrix}$ with odd determinant N and corresponding unimodular matrices $A=\frac{1}{\sqrt{N}}Y$.

Then the number of primitive integral matrices Y with

(V.13.2) $$A\in\Omega$$

possesses the asymptotic expression

(V.13.3) $$f(\Omega,N)\sim\Psi(N)\operatorname{meas}(\Omega)$$

as $N\to\infty$, where $\operatorname{meas}(\Omega)$ is the invariant measure of Ω on the group of unimodular matrices, defined by the formula (V.2.15), and $\Psi(N)$ is defined by

$$\Psi(N) = \frac{6}{\pi^2} V_0(N),$$

$$V_0(N) = \sum_{r^2|N} \mu(r) V_1\left(\frac{N}{r^2}\right); \quad V_1(N) = \sum_{\substack{n|N \\ n \leqq (\ln N)^{50}}} \frac{N}{n};$$

and

$$V_0(N) > \tfrac{1}{4} V_1(N).$$

We shall also need a small modification of Lemma V.11.1.

Lemma V.13.1: *Consider under the conditions of Lemma V.11.1 primitive matrices Y with* $\det(Y) = N = p^m$ *which are left divisible by a fixed matrix P with* $\det(P) = p$. *The number of such matrices $Y = PY'$ satisfying the condition $\frac{1}{\sqrt{N}} Y \in \Omega$ has the asymptotic expression*

$$f(\Omega, N, P) \sim \frac{\Psi(N)}{p+1} \operatorname{meas}(\Omega). \tag{V.13.4}$$

Proof: Let

$$Y = PY' \in \sqrt{N}\,\Omega; \quad P = \begin{pmatrix} p_1 & p_2 \\ p_3 & p_4 \end{pmatrix}.$$

Then

$$\bar{P}Y = pY' \in \sqrt{N} p \frac{\bar{P}}{\sqrt{p}} \Omega, \quad \text{where} \quad \frac{\bar{P}}{\sqrt{p}} \Omega$$

denotes the region on the unimodular group obtained from Ω through left multiplication by $\frac{\bar{P}}{\sqrt{p}}$.

We have $Y' \in \sqrt{\frac{N}{p}} \cdot \frac{\bar{P}}{\sqrt{p}} \Omega$.

Because the measure is invariant, $\operatorname{meas}\left(\frac{\bar{P}}{\sqrt{p}} \Omega\right) = \operatorname{meas}(\Omega)$. It is necessary to note, however, that Y' is not only primitive, but also not left divisible by $\bar{P}$, since PY' is primitive.

Further, from the definition of $\Psi(N)$ we easily find that

$$\Psi(N) = \Psi(p^m) = \frac{6}{\pi^2} p^{m-1}(p+1)\left(1 + \frac{\theta}{\ln^{40} N}\right). \tag{V.13.5}$$

Setting $\frac{6}{\pi^2} \operatorname{meas}(\Omega) = \mu$, we conclude from the preceding discussion that

$$f(\Omega, N, P) \sim \mu\left(V_0\left(\frac{N}{p}\right) - \frac{1}{\mu} f\left(\Omega', \frac{N}{p}, \bar{P}\right)\right), \tag{V.13.6}$$

where $\operatorname{meas}(\Omega')=\operatorname{meas}(\Omega)$ (in the following, $\operatorname{meas}(\Omega'')=\operatorname{meas}(\Omega''')$ $=\operatorname{meas}(\Omega)$). From (V.13.6) we deduce further that

$$f(\Omega, N, P) \sim \mu V_0\left(\frac{N}{p}\right) - V_0\left(\frac{N}{p^2}\right) - \frac{1}{\mu} f\left(\Omega'', \frac{N}{p^2}, P\right).$$

Hence with the help of (V.13.5), extending this further, we convince ourselves that

$$f(\Omega, N, P) \sim \mu p^{m-1}.$$

This coincides with (V.13.4) because of (V.13.5).

§ 14. Asymptotic Geometry of Hyperbolic Rotations

We return now to § 10. We dissected the Lobachevskian quadrangle $A_0(K_1)$ on the hyperboloid H_0 into regions $\Lambda_1, \Lambda_2, \ldots, \Lambda_{t_1}$.

Let $H_0(\Lambda_m)$ $(m=1,2,\ldots,t_1)$ be the number of fundamental and admissible points in the quadrangles in H whose projections to H_0 are in the Λ_m's. Therefore, because of (V.10.1),

$$\text{(V.14.1)} \qquad \begin{aligned} H_0(\Lambda_1)+H_0(\Lambda_2)+\cdots+H_0(\Lambda_{t_1}) &= H_0(A_0(K_1)) \\ &= h(-D)\left(1+\frac{\theta c_3}{\sqrt{K_1}}\right). \end{aligned}$$

From the numbers $H_0(\Lambda_m)$ we select those for which

$$\text{(V.14.2)} \qquad H_0(\Lambda_m) > \frac{h(-D)}{\ln^2 D}.$$

The remaining, if there are any, are discarded. Let $H_0(\Lambda_m)$ satisfy (V.14.2) for a certain m and let $L_1, L_2, \ldots, L_{h_m}$ $(h_m = H_0(\Lambda_m))$ be the corresponding fundamental and admissible vector matrices; $L_i^2 = -D$.

Let $\zeta_0, \zeta_1, \ldots$ be small positive constants to be chosen successively in the following. We fix the number k and the power p^k; its exact role will be explained later. Choose s such that

$$\text{(V.14.3)} \qquad D^{\frac{1}{2}+\zeta_2} \leqq p^{ks} < p^k D^{\frac{1}{2}+\zeta_2}.$$

Here $\zeta_2 = 10\zeta_1$ and $\zeta_1 > 0$ is explained later.

The number ζ_0 will be an arbitrarily small fixed constant.

We choose l satisfying the condition $0 < l < p^{ks}$. Consider further the equations

$$\text{(V.14.4)} \qquad l + L_\alpha = R_{\alpha 1} R_{\alpha 2} \ldots R_{\alpha s} V_\alpha; \qquad \alpha = 1, 2, \ldots, h_m,$$

L_α as above, belonging to Λ_m, and $R_{\alpha 1}, \ldots, R_{\alpha s}$ primitive matrices with determinant p^k. We consider further the primitive matrices

$$\text{(V.14.5)} \qquad T_{\alpha 1} = R_{\alpha 1};\ T_{\alpha 2} = R_{\alpha 1} R_{\alpha 2};\ \ldots;\ T_{\alpha s} = R_{\alpha 1} \ldots R_{\alpha s}.$$

We assume the $R_{\alpha j}$'s were chosen so that $T_{\alpha j}^{-1} L_\alpha T_{\alpha j} = L'_\alpha$ is a fundamental vector matrix. To them correspond the unimodular matrices

$$A_{\alpha 1} = \frac{T_{\alpha 1}}{\sqrt{\det(T_{\alpha 1})}};\ A_{\alpha 2} = \frac{T_{\alpha 2}}{\sqrt{\det(T_{\alpha 2})}};\ \ldots;\ A_{\alpha s} = \frac{T_{\alpha s}}{\sqrt{\det(T_{\alpha s})}}. \tag{V.14.6}$$

The total number of these unimodular matrices for all equations in (V.14.4) (assuming repetitions) will be sh_m.

In the group of unimodular matlices A we consider the region $\Omega(\Lambda_m, \Sigma_0)$ of unimodular matrices A which transfer the vertex 0_m of the quadrangle Λ_m chosen in § 10 into the region Σ_0 (see § 10 and the formulation of Theorem V.2.1). Namely, if $M = \begin{pmatrix} x_3 & -x_1 \\ x_2 & -x_3 \end{pmatrix}$ is a vector matrix corresponding to the vertex $0_m(x_1, x_2, x_3)$, then we have $A^{-1}MA \in \Sigma_0$. We will assume that Σ_0 is contained in H_0. We introduce a small $\eta > 0$, which will be specified later, and divide the second indices j of the matrices $A_{\alpha j}$ $(j = 1, 2, \ldots, s)$ into two types:

Type I – Those indices j, for which among the matrices $A_{\alpha j}$ with $\alpha = 1, 2, \ldots, h_m$, the number of first indices α for which $A_{\alpha j} \in \Omega(\Lambda_m, \Sigma_0)$ lies between

$$(1 - \eta_0)\frac{\Lambda(\Sigma_0)}{\Lambda(\Delta_0)} h_m \quad \text{and} \quad (1 + \eta_0)\frac{\Lambda(\Sigma_0)}{\Lambda(\Delta_0)} h_m. \tag{V.14.7}$$

Type II – Those indices j for which this condition is not valid.

Now we introduce a small number $\eta_1 > 0$ which is fixed in the following.

If for sufficiently large values of D and for the region Λ_m the number of type II indices is less than $\eta_1 s$, then we shall regard the result as satisfactory for this region Λ_m. We shall now prove that this is always true. Suppose it is not. Then we have one or the two following cases: Case 1. There exist $s_1 \geqq \eta_1 (\frac{1}{2} s)$ indices j for which the number of $A_{\alpha j} \in \Omega(\Lambda_m, \Sigma_0)$ $(\alpha = 1, 2, \ldots, h_m)$ is greater than

$$(1 + \eta_0)\frac{\Lambda(\Sigma_0)}{\Lambda(\Delta_0)} h_m. \tag{V.14.8}$$

Case 2. There exist $s_1 \geqq \eta_1 (\frac{1}{2} s)$ indices j for which the number of $A_{\alpha j} \in \Omega(\Lambda_m, \Sigma_0)$ $(\alpha = 1, 2, \ldots, h_m)$ is less than

$$(1 - \eta_0)\frac{\Lambda(\Sigma_0)}{\Lambda(\Delta_0)} h_m. \tag{V.14.9}$$

We consider Case 1. Let $j_1, j_2, \ldots, j_{s_1}$ be the corresponding indices and $A_{\alpha j_1}, A_{\alpha j_2}, \ldots, A_{\alpha j_{s_1}}$ $(\alpha = 1, 2, \ldots, h_m)$ the respective matrices.

Among them, the number of operators $A_{\alpha j_n} \in \Omega(\Lambda_m, \Sigma_0)$ will be by assumption larger than

$$(1 + \eta_0)\frac{\Lambda(\Sigma_0)}{\Lambda(\Delta_0)} h_m s_1. \tag{V.14.10}$$

We now count this number in the matrix

(V.14.11) $$\|A_{\alpha j_n}\|,$$

composed of the matrices $A_{\alpha j_n}$, first by rows and then by columns. Let h'_m be the number of rows for which $A_{\alpha j_n} \in \Omega(\Lambda_m, \Sigma_0)$ occurs more than $\left(1+\frac{\eta_0}{2}\right)\frac{\Lambda(\Sigma_0)}{\Lambda(\Delta_0)}s_1$ times, for fixed α and $j_n = j_1, \ldots, j_{s_1}$. Then we have

$$(h_m - h'_m)\left(1+\frac{\eta_0}{2}\right)\frac{\Lambda(\Sigma_0)}{\Lambda(\Delta_0)}s_1 + h'_m s_1 > (1-\eta_0)\frac{\Lambda(\Sigma_0)}{\Lambda(\Delta_0)}h_m s_1$$

hence

(V.14.12) $$h'_m > \frac{\eta_0}{2}\,\frac{\Lambda(\Sigma_0)}{\Lambda(\Delta_0)}h_m = \eta_2 h_m .$$

Here $\eta_2 = \frac{\eta_0}{2}\,\frac{\Lambda(\Sigma_0)}{\Lambda(\Delta_0)}$; in the following $\eta_2, \eta_3, \ldots$ will be small positive constants, chosen successively. Now we turn to Case 2. Let h''_m be the number of rows of (V.14.11) in which $A_{\alpha j} \in \Omega(\Lambda_m, \Sigma_0)$ occurs less than $\left(1-\frac{\eta_0}{2}\right)\frac{\Lambda(\Sigma_0)}{\Lambda(\Delta_0)}s_1$ times. Then we have

$$(1-\eta_0)\frac{\Lambda(\Sigma_0)}{\Lambda(\Delta_0)}h_m s_1 > (h_m - h''_m)\left(1-\frac{\eta_0}{2}\right)\frac{\Lambda(\Sigma_0)}{\Lambda(\Delta_0)}s_1$$

and

(V.14.13) $$h''_m > \frac{\eta_0}{2-\eta_0}h_m = \eta_3 h_m .$$

Therefore, in each of the above-mentioned cases we find no less than $\eta_4 h_m (\eta_4 = \min(\eta_2, \eta_3))$ first indices α, for which the number r_α of operators $A_{\alpha j}$ $(j = j_1, \ldots, j_{s_1})$ satisfying the condition $A_{\alpha j} \in \Omega(\Lambda_m, \Sigma_0)$ does not fulfill the inequalities

(V.14.14) $$(1-\eta_0)\frac{\Lambda(\Sigma_0)}{\Lambda(\Delta_0)}s_1 \leqq r_\alpha \leqq (1+\eta_0)\frac{\Lambda(\Sigma_0)}{\Lambda(\Delta_0)}s_1 .$$

We shall reduce this hypothesis to a contradiction.

Consider the $h''_m > \eta_4 h_m$ equations for h''_m – as described above:

(V.14.15) $$l + L_\alpha = R_{\alpha 1} R_{\alpha 2} \ldots R_{\alpha s} V_\alpha ; \qquad \alpha = 1, 2, \ldots, h''_m .$$

(with a rearrangement in numbering). Here $T_{\alpha j} = R_{\alpha 1} \ldots R_{\alpha j}$ is a primitive matrix. We select the special matrices $T_{\alpha j_1}, \ldots, T_{\alpha j_{s_1}}$: and the unimodular matrices $A_{\alpha j_1}, \ldots, A_{\alpha j_{s_1}}$ corresponding to them. If we have $A_{\alpha j_n} \in \Omega(\Lambda_m, \Sigma_0)$, then, in view of the condition that Σ_0 is contained in Δ_0 for sufficiently small grids on the region Λ_m (for sufficiently large numbers t_1, see § 10), the vector matrix

(V.14.16) $$T_{\alpha j_n}^{-1} L_\alpha T_{\alpha j_n} = L'_\alpha$$

will be fundamental. We note that the partition on the region Λ_m must be smaller than the smallest distance of Σ_0 from the boundary of Δ_0, measured in the Lobachevskian sense, but in the following we are free from this, taking also regions Σ_0 with points on the boundary of Δ_0.

We consider now primitive matrices Y_{j_1} with $\det(Y_{j_1})=p^{kj_1}$ and their respective

$$A_{j_1}=\frac{Y_{j_1}}{\sqrt{\det(Y_{j_1})}}.$$

Denote as before the vector matrix corresponding to the vertex $0_m(x_1,x_2,x_3)$ by

$$M=\begin{pmatrix} x_3 & -x_1 \\ x_2 & -x_3 \end{pmatrix}.$$

We consider the action of the matrices A_{j_1} on M in the sense of the results of the transformations $A_{j_1}^{-1}MA_{j_1}$. First we take into account all matrices Y_{j_1} for which $A_{j_1}^{-1}MA_{j_1}\in\Delta_0$. According to Lemma V.11.1, for any fixed large constant K' and the region $A_0(K')$ (see § 2), the number of Y_{j_1} for which

(V.14.17) $$A_{j_1}^{-1}MA_{j_1}\in A_0(K')$$

is given by the expression

(V.14.18) $$f(A_0(K'),p^{kj_1})=\Psi(p^{kj_1})\,\mathrm{meas}\,\Omega(\Lambda_m,A_0(K'))(1+\theta\eta_5(k)).$$

Here $\eta_5(k)\to 0$ for $k\to\infty$ and fixed K'. Since (see § 2)

$$\frac{\mathrm{meas}\,\Omega(\Lambda_m,A_0(K'))}{\mathrm{meas}\,\Omega(\Lambda_m,\Delta_0)}=\frac{\Lambda(A_0(K'))}{\Lambda(\Delta_0)},$$

we have

$$\mathrm{meas}\,\Omega(\Lambda_m,A_0(K'))=\mathrm{meas}\,\Omega(\Lambda_m,\Delta_0)\,(1+\theta\eta_6(K'));\ \eta_6(K')\to 0 \quad \text{for } K'\to\infty.$$

Hence, taking into account (V.13.5), we find that

(V.14.19) $$f(\Delta_0,p^{kj_1})=\frac{6}{\pi^2}\,\mathrm{meas}\,\Omega(\Lambda_m,\Delta_0)\,p^{kj_1-1}\times (p+1)\,(1+\theta\eta_7(k,K')).$$

We remark yet that if Y_{j_1} is a primitive matrix with $\det(Y_{j_1})=p^{kj_1}$, then, among the matrices Y_{j_1} right associated with Y_{j_1}, there exist exactly two: Y'_{j_1} and $Y''_{j_1}=-Y'_{j_1}$, which transform 0_m into Δ_0 according to the formula $Y_{j_1}^{-1}MY_{j_1}\in\Delta_0$.

This follows from the definition of the fundamental region Δ_0. (It is obtained by the reduction of positive non-integral forms with the help of the group of integral unimodular matrices ε.)

According to § 2 Chapter II, there exist exactly $p^{kj_1-1}(p+1)$ sets of right associated matrices $\{Y_{j_1}\varepsilon\}$.

In view of the above statements we conclude that

(V.14.20) $$f(\Delta_0,p^{kj_1})\sim 2p^{kj_1-1}(p+1).$$

Comparing (V.14.19) and (V.14.20), we obtain

$$\frac{6}{\pi^2}\operatorname{meas}\Omega(\Lambda_m, \Delta_0) = 2. \tag{V.14.21}$$

We now consider right non-associated primitive matrices Y_{j_1} for which $A_{j_1} \in \Omega(\Lambda_m, \Sigma_0)$. The number of them, according to Lemma V.11.1 and the reasoning just finished, will be

$$\frac{1}{2} f(\Omega(\Lambda_m, \Sigma_0), p^{kj_1}) \sim \frac{1}{2}\cdot\frac{6}{\pi^2}\operatorname{meas}\Omega(\Lambda_m, \Sigma_0)\, p^{kj_1-1}(p+1).$$

In view of this we can write:

$$\frac{1}{2} f(\Omega(\Lambda_m, \Sigma_0), p^{kj_1}) \sim \frac{\operatorname{meas}\Omega(\Lambda_m, \Sigma_0)}{\operatorname{meas}\Omega(\Lambda_m, \Delta_0)} W(p^{kj_1}), \tag{V.14.22}$$

where $W(p^{kj_1}) = p^{kj_1}(p+1)$ is the total number of sets of right associated matrices.

On the basis of § 2 we obtain

$$\frac{\operatorname{meas}\Omega(\Lambda_m, \Sigma_0)}{\operatorname{meas}\Omega(\Lambda_m, \Delta_0)} = \frac{\Lambda(\Sigma_0)}{\Lambda(\Delta_0)} = u \tag{V.14.23}$$

where $u \in (0, 1)$ is a positive constant. Of course, u is the same for all values of m.

§ 15. Evaluation of Probabilities

Let us now fix the matrix $T_{\alpha j_1}$ in the equations (V.14.15). We pass over to the matrix $T_{\alpha j_2}$. What is the number of right non-associated $T_{\alpha j_2}$ which may occur in the equations (V.14.15) and for which $A_{\alpha j_2} \in \Omega(\Lambda_m, \Sigma_0)$?

Here $S_{\alpha j_2} = T_{\alpha j_1}^{-1} T_{\alpha j_2}$ must be primitive and not right divisible by $P_{\alpha j_2}$ with determinant p. We have

$$S_{\alpha j_2} p^{-\frac{j_2-j_1}{2}} \in \Omega'(\Lambda_m, \Sigma_0) = \Omega'; \quad \Lambda(\Omega') = \Lambda(\Omega).$$

In view of this the quantity we seek will be given by Lemma V.13.1:

$$\begin{aligned} &\frac{1}{2}[f(\Omega', p^{(j_2-j_1)k}) - f(\Omega', p^{(j_2-j_1)k}, P_{\alpha j_1})] \\ &= \frac{1}{2}\cdot\frac{6}{\pi^2}\operatorname{meas}(\Omega')\, p^{(j_2-j_1)k}(1+\theta\eta_7(k, K')) \\ &= u p^{(j_2-j_1)k}(1+\theta\eta_7(k, K')). \end{aligned} \tag{V.15.1}$$

The number of those $T_{\alpha j_2}$ for which $A_{\alpha j_2} \notin \Omega(\Lambda_m, \Sigma_0)$ and which occur in our equations will obviously be

$$(1-u)\, p^{(j_2-j_1)k}(1+\theta\eta_7(k, K')). \tag{V.15.2}$$

We go further to the matrix $T_{\alpha j_3}$. For fixed $T_{\alpha j_2}$ and with exactly the same calculations we convince ourselves that the number of possible (but, obviously, not obligatory) values for the matrix $T_{\alpha j_3}$, which may occur in (V.14.15) and for which $A_{\alpha j_3} \in \Omega(\Lambda_m, \Sigma_0)$, will be

$$u p^{(j_3-j_2)k}(1+\theta\eta_7(k,K')).$$

We can extend this argument to $j_4, j_5, \ldots, j_{s_1}$. We consider $W(p^{kj_{s_1}}) = p^{kj_{s_1}-1}(p+1)$ possible values of the matrix $T_{\alpha j_{s_1}}$. For each such value $T_{\alpha j_{s_1}}$ which in reality occurs in the equations (V.14.15), we select the respective right non-associated matrices $T_{\alpha j_1}, \ldots, T_{\alpha j_{s_1}}$. If for an arbitrary matrix $T_{\alpha j_\beta}$, we have $A_{\alpha j_\beta} \in \Omega(\Lambda_m, \Sigma_0)$, then we shall say that "the event Ω happens inside of $T_{\alpha j_{s_1}}$".

The maximal number of different $T_{\alpha j_s}$, inside of which the event Ω happens exactly $r \leqq s_1$ times, according to the preceding, will have the asymptotic expression

(V.15.3)
$$W(p^{kj_{s_1}})\, C_{s_1}^r u^r (1-u)^{s_1-r}(1+\theta_1\eta_7)\ldots(1+\theta_{s_1}\eta_7);$$
$$|\theta_i| \leqq 1 \quad (i=1,2,\ldots,s_1); \quad \eta_7=\eta_7(k,K').$$

But the equations (V.14.15) were constructed such that (see § 14)

(V.15.4)
$$r>(1+\eta_0)us_1 \quad \text{or} \quad r<(1-\eta_0)us_1$$

simultaneously for all terms of the equation ($\alpha=1,2,\ldots,h_m$).

The factor $C_{s_1}^r u^r(1-u)^{s_1-r}$ has in part a probabilistic meaning; it is the probability of occurrence of an event exactly r times in a Bernoulli scheme with s_1 trials, if the probability of a single occurrence is u. Here the mathematical expectation of the number of occurrences of the event is $E(r)=us_1$, and the variance is given by $\sigma(r)=\sqrt{u(1-u)s_1}$. Hence we see that the inequalities (V.15.4) correspond to an anomalously large deviation of the number of occurrences of the event Ω from its mathematical expectation us_1. The probabilities for such large deviations are very small and may be estimated with the help of Theorem I.3.2. The corresponding estimate for the conditions (V.15.4) gives

(V.15.5)
$$\sum C_{s_1}^r u^r(1-u)^{s_1-r} < \exp(-\eta_8 s_1),$$

where $\eta_8=\eta_8(\eta_0)>0$ depends only on η_0.

We return to (V.15.3). Then,

(V.15.6)
$$(1+\theta_1\eta_7)\ldots(1+\theta_{s_1}\eta_7)<(1+\eta_7)^{s_1}<\exp(2s_1\eta_7(k,K')).$$

We shall assume that K' and k were chosen such that

$$2\eta_7(k,K')<\tfrac{1}{4}\eta_8.$$

Then the number w_1 of possible different values of the matrix $T_{\alpha j_{s_1}}$ ($\alpha=1,2,\ldots,h_m'''$) in the equations (V.14.15) receives the estimate

(V.15.7)
$$w_1 \leqq W(p^{kj_{s_1}})\exp(-\tfrac{3}{4}\eta_8 s_1).$$

For each given value of the matrix $T_{\alpha j_{s_1}}$ the integral matrix $T_{\alpha s} = R_{\alpha 1}\ldots R_{\alpha s}$ can take on no more than $p^{k(s-j_{s_1})}$ different right non-associated values, so that the total number of different $T_{\alpha s}$ in the equations

(V.14.15) does not exceed

$$(p+1)p^{ks}\exp(-\tfrac{3}{4}\eta_8 s_1)=(p+1)p^{ks-\frac{3}{4\ln p}\eta_8 s_1}.$$

According to § 14, $s_1 \geqq \eta_1 s/2$. Setting

$$\eta_9=\frac{\eta_1\eta_8}{4k\ln p}, \tag{V.15.8}$$

we find for sufficiently large s a lower estimate for the number w of possible different $T_{\alpha s}$:

$$w<p^{ks(1-\eta_9)}. \tag{V.15.9}$$

We turn to the basic lemma of § 3. We have, according to § 14, the number of equations (V.14.15) as $h_m''' > \eta_4 h_m = \eta_4(\eta_0)\, H_0(\Lambda_m)$. By virtue of (V.14.2) we get

$$h_m''' \geqq \frac{\eta_4(\eta_0)}{\ln^2 D}\, h(-D). \tag{V.15.10}$$

If the number η_9 is already chosen, then we can set $\zeta_1=\eta_9/12$; $\zeta_2=10\zeta_1$ and apply the basic lemma in § 3, substituting ks for s, η_9 for η_3, ζ_1 for η_1 and ζ_2 for η_2 (see § 14).

It is easy to see that for sufficiently large D we have arrived at a contradiction.

Therefore, for all of the regions Λ_m fulfilling the condition of "sufficient habitation" (V.14.2), the number of Type II indices will be less than $\eta_1 s$. The number of regions Λ_m is t_1; let the number of "sufficiently inhabited" regions in the sense of (V.14.2) be t_2. We form for each such Λ_m the equations of the form (V.14.4). In each of these t_2 systems of equations there will be less than $\eta_1 s$ indices j of Type II, and for all systems less than $t_2\eta_1 s \leqq t_1\eta_1 s$ Type II indices.

We set

$$\eta_1<1/t_1^2. \tag{V.15.11}$$

The the total number of indices of Type II will be $\leqq s/t_1$.

This means that there are not less than $s(1-1/t_1)$ indices j which are of Type I for all t_2 regions Λ_m.

Let $j=\beta$ be a particular index of this type.

We take the region Λ_m and construct the equations of the form (V.14.4):

$$l+L_\alpha=R_{\alpha 1}R_{\alpha 2}\dots R_{\alpha s}V_\alpha; \qquad \alpha=1,2,\dots,h_m.$$

Choosing the matrices $T_{\alpha j\beta}=R_{\alpha 1}\dots R_{\alpha j\beta}$, we find that

$$l+L_\alpha=T_{\alpha j\beta}V_\alpha' \qquad (\alpha=1,2,\dots,h_m). \tag{V.15.12}$$

The matrices $T_{\alpha j\beta}$ are such that $T_{\alpha j\beta}^{-1}L_\alpha T_{\alpha j\beta}$ is a fundamental vector matrix. This determines $T_{\alpha j\beta}$ among its right associates up to its sign; we choose any fixed sign for the corresponding matrix.

We consider the fundamental admissible vector matrices

(V.15.13) $$T_{\alpha j\beta}^{-1} L_\alpha T_{\alpha j\beta} = L'_\alpha \qquad (\alpha = 1, 2, \ldots, h_m).$$

All the vector matrices L'_α are different, as was proved in § 8. The number of them is $h_m = H_0(\Lambda_m)$.

Consider those matrices $T_{\alpha j\beta}$, actually occurring in the equations (V.15.12), for which

$$A_{\alpha j\beta} \in \Omega(\Lambda_m, \Sigma_0).$$

To them will correspond the integral fundamental vector matrices $T_{\alpha j\beta}^{-1} L_\alpha T_{\alpha j\beta}$. We remark that the statement in § 14 (concerning the arrangement of Σ_0 and Δ_0) is essential here.

By definition of the Type I indices (§ 14), the number of values α for which

(V.15.14) $$A_{\alpha j\beta} \in \Omega(\Lambda_m, \Sigma_0)$$

lies between the numbers in (V.14.7), that is, between the numbers

(V.15.15) $$(1-\eta_0) u h_m \quad \text{and} \quad (1+\eta_0) u h_m.$$

By virtue of the material in § 10 (see V.10.4) each of the vector matrices $T_{\alpha j\beta}^{-1} L_\alpha T_{\alpha j\beta}$ under the condition (V.15.14) will possess an image on H_0 lying in an expanded region $\Sigma'_0 \supseteq \Sigma_0$, for which (see V.10.4)

(V.15.16) $$1 \leqq \frac{\Lambda(\Sigma'_0)}{\Lambda(\Sigma_0)} \leqq 1 + \alpha(t_1),$$

where $\alpha(t_1) \to 0$ as $t_1 \to \infty$ uniformly in m. Therefore, we obtain the number of vector matrices with images in Σ_0 contained in the interval defined by (V.15.15).

Collecting for $m = 1, 2, \ldots, t_2$ all such vector matrices and taking into account that for all values α the $L'_\alpha = T_{\alpha j\beta}^{-1} L_\alpha T_{\alpha j\beta}$ are different, we get

(V.15.17) $$H_0(\Sigma'_0) \geqq (1-\eta_0) u \sum_{m=1}^{t_2} H_0(\Lambda_m).$$

Taking into account that $\sum_{m=t_2+1}^{t_1} H_0(\Lambda_m) = 0\left(\frac{h(-D)}{\ln^2 D}\right)$ in accord with the violation of condition (V.14.2), we obtain for large D

(V.15.18) $$H_0(\Sigma'_0) \geqq (1-2\eta_0) u H_0(\Lambda_0(K_1)).$$

Now we consider those $T_{\alpha j\beta}$ for which

(V.15.19) $$A_{\alpha j\beta} \notin \Omega(\Lambda_m, \Sigma_0).$$

According to (V.15.15) there are not less than

(V.15.20) $$h_m(1-(1+\eta_0)u)$$

such $T_{\alpha j\beta}$ for given m.

The action of the matrix $A_{\alpha j\beta}$ on the vertex 0_m of the region Λ_m carries 0_m out of the region $\Omega(\Lambda_m \cdot \Sigma_0)$. In view of this, the images on H_0

of the vector matrices $L'_\alpha = T^{-1}_{\alpha j\beta} L_\alpha T_{\alpha j\beta}$ will lie now outside of a region $\Sigma''_0 \subseteqq \Sigma_0$, chosen so that the distance between the boundaries of Σ''_0 and Σ_0, taken in the Lobachevskian sense, converge to zero for $t \to \infty$ and

$$1 \geqq \frac{\Lambda(\Sigma''_0)}{\Lambda(\Sigma_0)} > 1 - \alpha_1(t_1), \tag{V.15.21}$$

where $\alpha_1(t_1) \to 0$ as $t_1 \to \infty$.

Applying this discussion to all Λ_m, we obtain

$$\begin{aligned} H_0(\Sigma''_0) &\leqq \sum_{m \leqq t_1} h_m - h_m(1-(1+\eta_0)u) + 0\left(\frac{h(-D)}{\ln^2 D}\right) \leqq \\ &\leqq (1+2\eta_0)u H_0(\Lambda_0(K_1)). \end{aligned} \tag{V.15.22}$$

§ 16. Proof of Theorem V.2.1

Using the inequalities (V.15.18) and (V.15.22), it is not hard to prove Theorem V.2.1, if we make a suitable choice of the constants η_i and ζ_i. The inequalities (V.15.18) and (V.15.22) were proved for any Σ_0 with piecewise smooth boundary, not necessarily simply connected, lying in Δ_0. Here, nevertheless, on the basis of the considerations for satisfaction of the inequalities (V.15.18) and (V.15.22), it is necessary to choose D larger and larger, the nearer the boundary of Σ_0 comes to the boundary of Δ_0. (We remove this obstruction by chosing a new "almost" fundamental region Δ'_0 near to Δ_0 and containing Σ_0 in its interior.)

We apply the inequality (V.15.22) to the region $\Sigma'_0 - \Sigma''_0$. On the basis of (V.15.21) we get $\Lambda(\Sigma'_0 - \Sigma''_0) = \alpha_2(t_1) \to 0$ for $t_1 \to \infty$. Hence,

$$H_0(\Sigma'_0 - \Sigma''_0) \leqq (1+\eta_0)\frac{\alpha_2(t_1)}{\Lambda(\Delta_0)} h(-D).$$

We see that the number of fundamental admissible vector matrices with images on H_0 in this narrow strip will be as small as desired compared to $h(-D)$, if $D \to \infty$ and the Lobachevskian area of the strip is sufficiently small. This proves finally Theorem V.2.1:

$$H_0(\Sigma_0) = \frac{\Lambda(\Sigma_0)}{\Lambda(\Delta_0)} h(-D)\,(1+\eta(p, K_1, D)).$$

It remains only to justify the choice of the constants η_i and ζ_j. If we want $|\eta(p, K_1, D)| < \zeta_0$ (see § 14), then we choose first of all K_1 large enough, so that the ratio $\dfrac{\Lambda(\Lambda_0(K_1))}{\Lambda(\Delta_0)}$ is sufficiently small compared to 1. Then we choose the number t large enough (the partition small enough), so that the numbers $\alpha_1(t_1)$ (V.15.21) and $\alpha(t_1)$ (V.10.4) will be sufficiently small. After this a sufficiently small η_0 is chosen (see § 14). The number η_1 is assumed to be equal to $\dfrac{1}{2t_1^2}$ (see (V.15.11)). The number $\eta_8 = \eta_8(\eta_0)$

is selected according to η_0. Finally we choose k, K' so large, that $2\eta_7(K', k) < \frac{1}{4}\eta_8$ (see (V.15.7)). The number k is fixed. Then

$$\eta_9 = \frac{\eta_1 \eta_8}{4k \ln p}$$

and ζ_1, ζ_2 (see § 15) are defined. For sufficiently large D all considerations are valid. This completes the proof of Theorem (V.2.1). It can be written in the form

$$H(\Sigma) = \frac{9}{2\pi} \Lambda(\Sigma_0)\, h(-D)\,(1 + \eta(p, K_1, D)). \tag{V.16.1}$$

§ 17. Proofs of Theorems V.2.2 and V.2.3

We turn to the proof of Theorem V.2.2. Of course, it follows from Theorem V.2.1. Let Σ_0 be the region in Δ_0 for which $x_1 \leqq \alpha$. If $\alpha \geqq \sqrt{\frac{4}{3}}$, then $\Sigma_0 = \Delta_0$ and consequently the third statement in Theorem V.2.2 is trivial. Now we make use of the interpretation of Lobachevskian area as the Euclidean area of the cone based on Σ_0 (see § 2). We get

$$\Lambda(\Sigma_0) = 2 \iiint_{\Omega} dx_1\, dx_2\, dx_3\,, \tag{V.17.1}$$

where the region Ω is defined by the inequalities

$$x_2 \geqq x_1 \geqq 0; \quad 0 \leqq x_3 \leqq x_1/2; \quad x_1 \leqq \alpha; \quad 0 \leqq x_1 x_2 - x_3^2 \leqq 1.$$

We introduce the substitution $x_1 x_2 - x_3^2 = u$; $x_2 = \dfrac{u + x_3^2}{x_1}$ and go over to the variables (x_1, x_3, u). We obtain

$$\Lambda(\Sigma_0) = 2 \iiint_{\Omega_1} \frac{dx_1\, dx_3\, du}{x_1},$$

Ω_1 being defined by the inequalities $x_1 \leqq \alpha$; $0 \leqq x_3 \leqq x_1/2$; $0 \leqq x_1 x_2 - x_3^2 \leqq 1$. Further we set $x_3/x_1 = \xi$. Then,

$$\Lambda(\Sigma_0) = 2 \iiint_{\Omega_2} dx_1\, d\xi\, du\,,$$

Ω_2 being determined by the inequalities

$$0 \leqq \xi \leqq \tfrac{1}{2}; \quad 0 \leqq x_1 \leqq \alpha; \quad x_1^2(1 - \xi^2) \leqq u \leqq 1.$$

Finally we put $z = \dfrac{x_1}{\sqrt{u}}$ and obtain

$$\Lambda(\Sigma_0) = 2 \iiint_{\Omega_3} \sqrt{u}\, du\, d\xi\, dz$$

with the region Ω_3:

$$0 \leqq \xi \leqq \frac{1}{2}; \quad 0 \leqq u \leqq 1; \quad 0 \leqq z \leqq \min\left(\alpha, \frac{1}{\sqrt{1 - \xi^2}}\right),$$

so that

$$\Lambda(\Sigma_0) = \tfrac{4}{3} \iint_{\Omega_4} d\xi\, dz$$

with the region Ω_4:

$$0 \leqq \xi \leqq \frac{1}{2}; \quad 0 \leqq z \leqq \min\left(\alpha, \frac{1}{\sqrt{1 - \xi^2}}\right).$$

If $\alpha \leqq 1$, then $\min\left(\alpha, \frac{1}{\sqrt{1-\xi^2}}\right) = \alpha$; if $1 \leqq \alpha \leqq \sqrt{\frac{4}{3}}$, the region decomposes for values of ξ into the regions $0 \leqq \xi \leqq \sqrt{1-1/\alpha^2}$ and $\sqrt{1-1/\alpha^2} \leqq \xi \leqq \frac{1}{2}$, and we obtain after an easy calculation

$$\Lambda(\Sigma_0) = \tfrac{4}{3} \arcsin \sqrt{1-1/\alpha^2} + \tfrac{2}{3}\alpha\left(1-2\sqrt{1-1/\alpha^2}\right).$$

Hence,

$$\frac{\Lambda(\Sigma_0)}{\Lambda(\Delta_0)} = f(\alpha) = \frac{6}{\pi} \arcsin \sqrt{1-1/\alpha^2} + \frac{3}{\pi}\alpha\left(1-2\sqrt{1-1/\alpha^2}\right),$$

which concludes the proof of Theorem V.2.2.

We pass over to Theorem V.2.3. This theorem, formulated in terms of characters and Dirichlet series, is a corollary of part of Theorem V.2.2.

Let the number a take on the values of first coefficients of properly primitive reduced binary forms with discriminant $(-D)$ (or $(-D/4)$ respectively). From the elementary theory of L-series it follows that two Dirichlet series

$$\sum a^{-s} \quad \frac{\zeta(s)\, L(s)}{\zeta(2s)},$$

for $a \leqq \varepsilon_0 \sqrt{D}$ and sufficiently small ε_0, have coincident first terms in the expansion. The same is true, consequently, for the series

$$\zeta(2s) \sum a^{-s} \quad \text{and} \quad \zeta(s)\, L(s) = \sum_{n=1}^{\infty} a_n n^{-s}$$

for $s > 1$. Further, for each fixed $\varepsilon \in [0, 1]$, according to Theorem V.2.2,

$$\sum_{\alpha \leqq \varepsilon\sqrt{D}} 1 \sim \frac{3\varepsilon}{\pi} h(-D).$$

It is known, moreover, that

$$L(1) = \frac{\pi h(-D)}{2\sqrt{D}}; \quad h(-D) = 2/\pi \sqrt{D}\, L(1). \tag{V.17.2}$$

From the above and $\zeta(r) = \pi^2/6$ it follows that

$$\sum_{n \leqq \varepsilon\sqrt{D}} a_n \sim \frac{\pi^2}{6} \sum_{a \leqq \varepsilon\sqrt{D}} 1 \sim \frac{\pi\varepsilon}{2} h(-D).$$

In view of (V.17.2) we obtain

$$\sum_{n \leqq \varepsilon\sqrt{D}} a_n \sim \varepsilon \sqrt{D}\, L(1),$$

which we were trying to prove.

§ 18. On Ergodic Theorems for the Flow of Primitive Points of the Hyperboloid of Two Sheets

The Ergodic Theorem V.2.4 and the mixing Theorem V.2.6 can be deduced from the previous considerations in a completely analogous manner to Chapter IV. A theorem entirely analogous to Theorem III.1.5 is valid

Theorem V.18.1: *Let the properly primitive form* (a, b, c) *govern the hyperbolic rotation* (L, L_1) *according to the equations*

$$b+L=AC; \quad CLC^{-1}=L_1; \quad \det(A)=a; \quad \det(C)=c.$$

Further, suppose that the properly primitive form (a_1, b_1, c_1) *governs the rotation* (L_1, L_2) *by the equations*

$$b_1+L_1=A_1C_1; \quad C_1L_1C_1^{-1}=L_2; \quad \det(A_1)=a_1; \quad \det(C_1)=c_1.$$

Then the rotation (L, L_2) *will be governed by the class of forms* (a_2, b_2, c_2) *which is the Gaussian composition of the classes corresponding to the forms* (a, b, c) *and* (a_1, b_1, c_1).

The proof of this theorem is completely analogous to the proof of Theorem III.1.5 given in § 1 of Chapter III.

We see in this way that the flow of fundamental primitive points on the hyperboloid can be constructed from distinct cycles; trajectories, obtained from initial points by consecutive applications of the class of forms (p). A cycle containing the point corresponding to the class (a), for example, will consist of points corresponding to (a), $(a)(p)$, $(a)(p)^2$, $\ldots, (a)(p)^k, \ldots$; from the Ergodic Theorem V.2.4 we can draw the following conclusion: If we consider on cycles of the above type segments of length $s > c_0 \ln D$ and points corresponding to these segments, then for an overwhelming majority of such segments their points will be uniformly distributed on the fundamental region of the hyperboloid. Hence it is possible to deduce again Theorem V.2.1.

We consider the flow of primitive points obtained by the action of the whole group of classes H. If it is known that H contains a subgroup G with index g bounded by the constant c, then Theorem V.2.5, proved in exactly the same manner as Theorem V.2.4, shows us that any coset $G_A = A \cdot G$ of the subgroup G results also in an asymptotic uniform distribution in the Lobachevskian sense. Further, let us introduce the transformation T corresponding to the composition with the class $(p)^r$. Then the coset G_A will be invariant under the transformation: $TG_A = G_A$. We take this coset as the set N in the mixing Theorem V.2.6. From (V.2.8) it follows immediately that for any region Σ satisfying the conditions described in § 1 and lying in the fundamental region of the hyperboloid, the number of points of representatives of the coset G_A will be

$$M(G_A, \Sigma) = \frac{h(-D)}{g} \Lambda(\Sigma)\,(1+o(1))$$

as $D \to \infty$.

§ 19. Ergodic Theorems for a Modular Invariant

As is known, the theory of positive binary quadratic forms is closely connected with the classical modular invariant $I(\omega)$ and with the modular figure (see WEBER). The form (a, b, c) corresponds to the root $\omega = \frac{-b+\sqrt{D}}{2a}$; reduced forms correspond to roots ω lying in the fundamental region Δ_0 of values $z = x + iy$ for which

$$-\tfrac{1}{2} \leqq x \leqq \tfrac{1}{2}, \quad x^2 + y^2 \geqq 1 .$$

The modular figure is obtained from this region by transformation of the modular group. For given $-D$ each class of forms (a, b, c) in Δ_0 has exactly one representation.

If we measure area in Δ_0 with the help of the Poincaré model of the Lobachevskian plane, then the Ergodic Theorem V.2.4 and the mixing Theorem V.2.5 can be paraphrased in terms of points $\omega \in \Delta_0$, representing roots of reduced forms (a, b, c). However, completely new interpretations are possible. We consider two of these.

We defined the flow transformation T earlier with the help of composition with the form $(p, \xi, n) = P$. Following WEBER § 118, we introduce the exponent δ for which the class P^δ becomes principal. If (a) is any class of the group H, then for P and (a) representatives $(p, b, acp^{\delta-1})$ and (a, b, cp^δ) may be chosen, and for the class $(a)P$ the representative $(ap, b, p^{\delta-1}c)$. If ω is the root of the representation of (a), then ω/p will be the root of the representation of $(a)P$, ω/p^2 the root of the representation, etc., until $\omega/p^{\delta-1}$. For fixed P and large D, we obviously have $\delta > c_1 \ln D$.

We consider the points $\omega, \omega/p, \omega/p^2, \ldots, \omega/p^r, \ldots$ on the modular group as a trajectory of the flow of representations of the classes of forms. We shall transform each such point in the fundamental region Δ_0 with transformations of the modular group. The points thus obtained give the old trajectory of the flow of classes in Δ_0, and for it the Ergodic Theorem V.2.4 and the mixing Theorem V.2.6 are valid. But instead of the class composition we now discuss the sequence of points $\omega, \omega/p, \omega/p^2, \ldots$ generating the segment of the ray passing through $z=0$ and $z=\omega$ and forming a geometric progression. The operation of transforming these points into the fundamental region Δ_0 is completely analogous to the determination of the "fractional part" of a complex number lying inside a quadratic lattice, if the modular figure is regarded as an analogy to such a lattice in the replacement of Euclidean geometry by Lobachevskian geometry. Therefore the analogies of the Ergodic Theorem V.2.4 and V.2.6 may be considered as theorems on the behavior of fractional parts of the geometric series $\omega, \omega/p, \omega/p^2, \ldots$, although they would

8*

apparently be very difficult to prove in view of the known peculiarities in behavior of the modular figure in the vicinity of the real axis.

For the second interpretation we make use of properties of the modular invariant $I(\omega)$ of the roots ω. It is known that for each class (a) and fixed P

(V.19.1) $$I(\omega(aP)) = f_p(I(\omega(a)), \sqrt{D})$$

(see WEBER § 121). Here $\omega(a)$, $\omega(aP)$ are the images of the classes (a) and (aP) in Δ_0, and f is a rational function depending on D and P. (A simplification is available for $p=2$.) Here the operation T of our flow (composition with the class P) can be reduced to composition with the rational function (V.19.1), and the values $I(\omega)$ for points on a trajectory are obtained by a simple iteration of the rational function (V.19.1). The Ergodic Theorems V.2.4 and V.2.6 will then become theorems on the iteration of the rational function (V.19.1). However, it is not at all easy to prove them in this form.

§ 20. Supplementary Remarks

The problem of the remainder in Theorem V.2.1 is completely analogous to the same problem for the distribution of integral points on the sphere (see § 8, Chapter IV). Using those methods and making more economical estimates it can easily be shown that in the formula (V.2.1) the remainder will satisfy the inequality

(V.20.1) $$\eta(p, K_1, D) < c(p, K_1)\,(\ln D)^{-\alpha},$$

where α is a positive constant. However, a further improvement of the remainder is very difficult.

A significant deficiency of method turns out to be the introduction of the auxiliary number p for which $(D/p) = +1$. This can be avoided if certain hypotheses on the zeroes of L-series are proved; in this case the proofs can be simplified, although the remainder is also very poor for this method. We shall consider problems of this type in Chapter IX.

We make a few remarks connected with Theorem V.2.3. Let $-D<0$ be a fundamental discriminant. Then for arbitrarily small fixed $\varepsilon>0$ we have

(V.20.2) $$\sum_{n \leqq \varepsilon\sqrt{D}} a_n \sim \varepsilon\sqrt{D}\, L(1, X),$$

where $X = X(n) = (-D/n)$ is a real character $(\mathrm{mod}\, D)$. The behavior of X as $D \to \infty$ is connected with the known hypothesis of I. M. VINOGRADOV (see, for instance, VINOGRADOV [2]). If $X(n) = -1$, then for some prime divisor q of n we have $X(q) = -1$. The smallest such prime number $q = q_D$ has been studied by many authors. According to one of VINOGRADOV's hypotheses, we have

$$\lim_{D\to\infty} \frac{\ln q_D}{\ln D} = 0.$$

I. M. VINOGRADOV proved in 1926 that

$$\lim_{D\to\infty} \frac{\ln q_D}{\ln D} \leqq \frac{1}{2\sqrt{e}}.$$

In 1958 D. BURGESS used a theorem of A. WEIL on the number of prime divisors for a curve in a finite field to show that

$$\lim_{D\to\infty} \frac{\ln q_D}{\ln D} \leqq \frac{1}{4\sqrt{e}}.$$

Further progress in this direction was not obtained. Another hypothesis due to I. M. VINOGRADOV deals with the smallest prime number $q = q'_D$ for which $X(q) = +1$. This hypothesis states that

$$\lim_{D\to\infty} \frac{\ln q'_D}{\ln D} = 0.$$

From a theorem of K. L. SIEGEL in 1935 (the relation (V.1.9)) it follows that $q'_D \leqq \sqrt{\frac{4}{3}D}$, hence

$$\lim_{D\to\infty} \frac{\ln q'_D}{\ln D} \leqq \frac{1}{2}.$$

Theorem V.2.3 (relation (V.20.2)) allows us to give a much better estimate. If such a $p \geqq 3$ exists with $X(p) = +1$ and if D is sufficiently large in comparison with p, then we have (V.20.2) (the asymptotic dependence on p is essential here). We fix a small $\varepsilon > 0$. Of course, the number p plays the role of q'_D. We consider, however, prime numbers $q \neq p$ for which $X(q) = +1$ and we look for the smallest number $q''_D \neq p$ among them. According to (V.20.2), there exist numbers n not exceeding $\varepsilon\sqrt{D}$ for which all prime divisors q enjoy the property $X(q) = +1$. These numbers n are not all powers of p, since the number of powers of p not exceeding $\varepsilon\sqrt{D}$ is of the order $0(\ln D)$, and the right side of (V.20.2) is not less than

$$c(\varepsilon')\, D^{\frac{1}{2}-\varepsilon'}$$

for any $\varepsilon' > 0$, according to SIEGEL's theorem (see (V.1.9)). Therefore there exists a prime number q, different from p, such that $q \leqq \varepsilon\sqrt{D}$ and $X(q) = +1$. Moreover, it can be shown that the number of such primes, using (V.20.2), is not bounded as $D \to \infty$. More economical estimates in the preceding proofs lead to the relation (V.20.1) by substituting $\frac{1}{(\ln D)^\alpha}$ for ε in our assertion, where $\alpha > 0$ is a constant. Thus, the ergodic methods yield little in this problem.

At the same time we see (see LINNIK-VINOGRADOV), that an application of A. WEIL's proof of the Riemann hypothesis for zeta functions of hyperelliptic curves on prime finite fields reduces to a significant advancement in the stated direction. Namely, the relation (V.20.2) is strengthened in the following manner: for any $\varepsilon > 0$, $x \geqq D^{\frac{1}{4}+\varepsilon}$ and $D > D_0(\varepsilon)$ we have

$$\sum_{n \leqq x} a_n \sim x L(1, X) \tag{V.20.3}$$

Hence it follows that the smallest prime number $q = q'_D$ with $X(q) = +1$ is of the order

$$q'_D = 0(D^{\frac{1}{4}+\varepsilon}) \tag{V.20.4}$$

for arbitrary $\varepsilon > 0$.

The relation (V.20.3) permits a considerable simplification of the proofs of the theorems in Chapters IV, V, and VI, and provides the proofs with a new fundament. We must limit ourselves to the above remarks.

Chapter VI

Flows on Primitive Points on a Hyperboloid of One Sheet

§ 1. Formulation of the Problem

In Chapters IV and V we considered flows of primitive points on the sphere and the hyperboloid of two sheets. In order to accomplish this, we studied solutions of the equations $L^2 = -m$ $(m > 0)$ for quaternions, that is, for special 4×4-matrices, and solutions of the equations $L^2 = -D (D > 0)$ for integral 2×2-matrices. The matrices mentioned gave us representations of the field $k(\sqrt{-m})$ or $k(\sqrt{-D})$, i.e., an imaginary quadratic field (concerning this, see Chapter VII). Among all algebraic fields the imaginary quadratic fields are characterized by the following property: only they have finitely many units.

With these fields are closely connected binary quadratic forms, admitting geometrically a simple characterization of their reduced regions. Another result is obtained by consideration of integral matrices satisfying the equation

$$L^2 = D \qquad (D > 0), \tag{VI.1.1}$$

where D is not a perfect square.

These matrices give us a representation of the real quadratic field $k(\sqrt{D})$. Such a field, as is known, has infinitely many units and properties closer to an algebraic expansion of the field of rational numbers than to the field $k(\sqrt{-D})$.

Therefore for the transition to flows and ergodic theorems connected with irreducible equations of n-th degree

$$f(L) = 0, \tag{VI.1.2}$$

where L is an $n \times n$-matrix, it is important to investigate the simplest case of equation (VI.1.1).

Such investigations are given in B. F. Skubenko [1]. We expound this work here.

§ 2. Formulation of the Basic Theorem. A Lemma on Integral Points

We consider the surface of the hyperboloid of one sheet H in cartesian coordinates a, b, c:

$$H : b^2 - ac = D, D > 0. \tag{VI.2.1}$$

The number D is assumed to be integral, odd, and not a perfect square. We choose the classical region of reduction Δ on the surface of H:

$$\Delta : 0 < b < \sqrt{D}, \ \sqrt{D} - b < |a| < \sqrt{D} + b, \ b^2 - ac = D,$$

and introduce the following notation: denote by Δ_1 a finitely connected figure with piecewise smooth boundary in Δ; by $V(\Delta_1)$ the volume of the cone with base at Δ_1 and vertex at the origin; by $A(\Delta_1)$ the hyperbolic solid angle

$$A(\Delta_1) = V(\Delta_1) \cdot D^{-\frac{3}{2}};$$

by $T(\Delta_1)$ the number of integral points lying in the surface Δ_1 for which $(a, 2b, c) = 1$.

Let there exist a prime number such that $(D/p) = +1$. Then the following theorem is valid.

Theorem VI.2.1: *As* $D \to \infty$

$$T(\Delta_1) = \frac{A(\Delta_1)}{A(\Delta)} T(\Delta)\,(1 + \eta(p, D)),$$

where $\eta(p, D) \to 0$ *for fixed* p *and* $A(\Delta_1)$.

Theorem VI.2.1 is proved with the same method which we used to prove the analogous theorem for the case of the hyperboloid of two sheets in Chapter V. Here the presence of non-trivial units in real fields leaves its imprint on all steps of the proof of Theorem VI.2.1. A basic difficulty in the proof is overcome by the introduction of the "theorem on cycles", which is of independent interest.

We shall identify the point (a, b, c) with the quadratic form

$$\varphi(x, y) = ax^2 + 2bxy + cy^2$$

and with the matrix

$$L = \begin{pmatrix} b & -a \\ c & -b \end{pmatrix},$$

and we shall assume that in the following these three objects always correspond to each other. On the hyperboloid (VI.2.1) we consider in particular the following regions:

$$\begin{aligned}
&\mathrm{I}: b^2 - ac = D,\ 0 < |b| < \sqrt{D},\ \sqrt{D} - |b| < |a| < \sqrt{D} + |b|;\\
&\mathrm{II}: b^2 - ac = D,\ 0 < b < \sqrt{D},\ \sqrt{D} - b < |a| < \sqrt{D} + b;\\
&\mathrm{III}: b^2 - ac = D,\ 0 < b < \sqrt{D},\ \sqrt{D} - b < |a| < \sqrt{D} + b,\ c > 0;\\
&\mathrm{IV}: b^2 - ac = D,\ 0 < b < \sqrt{D},\ \sqrt{D} - b < |a| < \sqrt{D} + b,\ c < 0.
\end{aligned}$$

Region II is the classical region of reduction; Regions III and IV are its composite parts.

We denote by ε a matrix with $\det \varepsilon = +1$ which has the form

$$\varepsilon = t - uL.$$

Then it is obvious that

$$\varepsilon^n L \varepsilon^{-n} = L,$$

where n is any integer.

Let $L_1, L_2, \ldots, L_{2k}$ be representatives of an arbitrary class of the Region II, numbered so that the matrix L_{i+1} is a right neighbor of L_i; further, let the matrices

$$\begin{pmatrix} 0 & -1 \\ 1 & z_1 \end{pmatrix}, \begin{pmatrix} 0 & -1 \\ 1 & z_2 \end{pmatrix}, \ldots, \begin{pmatrix} 0 & -1 \\ 1 & z_{2k} \end{pmatrix}$$

have the property

$$\begin{pmatrix} 0 & -1 \\ 1 & z_i \end{pmatrix} L_i \begin{pmatrix} 0 & -1 \\ 1 & z_i \end{pmatrix}^{-1} = L_{i+1}.$$

It is then clear that

$$\pm \prod_{i=1}^{2k} \begin{pmatrix} z_i & 1 \\ -1 & 0 \end{pmatrix} = T + U \cdot L_1,$$

where T and U are the smallest positive solutions of PELL's equation

$$T^2 - DU^2 = 1.$$

Let the congruence $x^2 \equiv D \pmod{N}$ be solvable, N odd, $A = \begin{pmatrix} \alpha & \beta \\ \gamma & \delta \end{pmatrix}$, and $\det(A) = N$. Then the relations

$$l + L = B \cdot A, \quad l + L' = A \cdot B$$

make sense, where l is integral, A, B, L are integral 2×2 matrices for which

$$L = \begin{pmatrix} b & -a \\ c & -b \end{pmatrix}, \quad (a, 2b, c) = 1,$$

$$L' = \begin{pmatrix} b' & -a' \\ c' & -b' \end{pmatrix}, \quad (a', 2b', c') = 1.$$

We denote by A_i the matrix $A\varepsilon_L^{-i}$, where

$$\varepsilon_L = T - UL.$$

Obviously

$$\text{(VI.2.2)} \qquad \begin{aligned} \alpha_i &= \alpha t - (b\alpha + c\beta)u, & \beta_i &= \beta t + (a\alpha + b\beta)u, \\ \gamma_i &= \gamma t - (b\gamma + c\delta)u, & \delta_i &= \delta t + (a\gamma + b\delta)u. \end{aligned}$$

Since $A_i L A_i^{-1} = L'$, then

$$\text{(VI.2.3)} \qquad \begin{aligned} a'N &= a\alpha_i^2 + 2b\alpha_i\beta_i + c\beta_i^2 \\ b'N &= a\alpha_i\gamma_i + b(\alpha_i\delta_i + \beta_i\gamma_i) + c\beta_i\delta_i \\ c'N &= a\gamma_i^2 + 2b\gamma_i\delta_i + c\delta_i^2. \end{aligned}$$

Let $aa'N > 0$ (as we shall see from the following, we may make this assumption with no loss of generality).

We shall prove that the following hypothesis can be imposed on A_i:

$$(*) \qquad (a\alpha_i + b\beta_i)\, U > T\beta_i \geqq 0.$$

Indeed, let $T + U\sqrt{D} = \theta$ and let the condition $(*)$ be fulfilled; then

$$(\theta - \theta^{-1})\,(a\alpha_i + b\beta_i) > (\theta + \theta^{-1})\,\beta_i \sqrt{D} \geqq 0,$$

or

$$\theta^2(a\alpha_i+b\beta_i)-\theta^2\beta_i D > a\alpha_i+b\beta_i+\beta_i\sqrt{D}>0,$$
$$\theta^2(a\alpha_i+(b-\sqrt{D})\beta_i) > a\alpha_i+(b+\sqrt{D})\beta_i \geqq$$
$$\geqq a\alpha_i+(b-\sqrt{D})\beta_i>0,$$
$$\theta(a\alpha_i+(b-\sqrt{D})\beta_i)^{\frac{1}{2}} > (a\alpha_i+(b+\sqrt{D})\beta_i)^{\frac{1}{2}} \geqq (a\alpha_i+(b-\sqrt{D})\beta_i)^{\frac{1}{2}}>0.$$

Multiplying the last formula by

$$(a\alpha_i+(b+\sqrt{D})\beta_i)^{\frac{1}{2}},$$

we get

$$\theta\sqrt{aa'N} > a\alpha_i(b+\sqrt{D})\beta_i \geqq \sqrt{aa'N}>0.$$

We substitute for α_i, β_i their values as in (VI.2.2); then we find that

$$\theta\sqrt{aa'N} > (t-u\sqrt{D})\,(a\alpha+(b+\sqrt{D})\beta) \geqq \sqrt{aa'N}$$

or

$$(**)\qquad \theta\sqrt{aa'N} > \pm\theta^k(a\alpha+(b+\sqrt{D})\beta) \geqq \sqrt{aa'N}>0.$$

Hence, we see that t and u are uniquely determined by the condition (∗). If the relation (∗) is proved, then it is always possible to attain the relation (∗∗) for suitable k. Performing the calculation in the opposite direction, we obtain the relation (∗).

In this manner we have deduced the following lemma.

Lemma VI.2.1:

1) *If the matrix* $A(\det A>0)$ *carries* L *into* L', *then the matrix* $A_i=A\cdot\varepsilon_L^{-i}$ *carries* L *into* L'.

2) *Among the matrices* $\pm A_i$ $(i=\pm1, \pm2, \ldots)$ *(for given* l*) there exists exactly one* A_i *such that*

$$A_i=\begin{pmatrix}\alpha_i & \beta_i\\ \gamma_i & \delta_i\end{pmatrix},\quad (a\alpha_i+b\beta_i)U>\beta_i T\geqq 0.$$

Let

$$L,\ L'\in \mathrm{IV},\ ALA^{-1}=L',\ (a\alpha+b\beta)U>\beta T\geqq 0,$$
$$A_i=\begin{pmatrix}\alpha & \beta\\ \gamma & \delta\end{pmatrix}\varepsilon_L^{-1};$$

then

$$A_i=\pm\varepsilon_{L'}^{-1}\begin{pmatrix}\alpha & \beta\\ \gamma & \delta\end{pmatrix}$$

$$(\mathrm{VI.2.4})\qquad \begin{aligned}\alpha_i&=\alpha t-(b\alpha+c\beta)u, & \beta_i&=\beta t+(a\alpha+b\beta)u,\\ \gamma_i&=\gamma t+(b\gamma+c\delta)u, & \delta_i&=\delta t+(a\alpha+b\beta)u,\\ \alpha_i&=\alpha t-(b'\alpha-a'\gamma)u, & \beta_i&=\beta t+(a'\delta-b'\beta)u,\\ \gamma_i&=\gamma t-(c'\alpha-b'\gamma)u, & \delta_i&=\delta t+(b'\delta-c'\beta)u.\end{aligned}$$

Since $a\alpha+b\beta=a'\delta-b'\beta$ and $|t|-b|u|>0$, then

$$\alpha>0,\quad \beta\geqq 0,\quad \delta>0.$$

Making use of (VI.2.3) and (VI.2.4), we obtain

a) if $t>0$, $u>0$, that is, $i>0$, then

(VI.2.5) $$\beta_i>\alpha_i>0, \quad \delta_i>\gamma_i>0;$$

b) if $t>0$, $u<0$, that is, $i<0$, then

(VI.2.6) $$\alpha_i>-\beta_i>0, \quad -\gamma_i>\delta_i>0.$$

for sufficiently large $|i|$. Hence we have the following lemma.

Lemma VI.2.2: *If $L, L' \in$ IV, $ALA^{-1}=L'$, $\det A>0$, then there exists an i such that $A\varepsilon_L^{-i}=A_i$ fulfills* (VI.2.5) *and $A\varepsilon_L^i=A_{-i}$ satisfies* (VI.2.6). *Obviously, $A_i=A_{-i}\varepsilon_L^{2i}$.*

Lemma VI.2.3: *If $L, L' \in$ II, $ALA^{-1}=L'$, $\det A=N>0$, then there exists a matrix $L''=\varepsilon L\varepsilon^{-1}$, where ε is a unimodular matrix, such that $L'' \in$ II, $A'=A\varepsilon^{-1}$ and $\alpha'\beta'\gamma'\delta' \leqq 0$.*

Proof: Let $L=L_1, L_2, \ldots, L_{2k}$ be all the representatives of a class on the Region II, numbered in such a way that the matrix L_{j+1} is a right neighbor of L_j, and let the matrices

$$\begin{pmatrix}0 & -1\\ 1 & z_1\end{pmatrix}, \begin{pmatrix}0 & -1\\ 1 & z_2\end{pmatrix}, \ldots, \begin{pmatrix}0 & -1\\ 1 & z_{2k}\end{pmatrix}$$

be chosen so that

$$\begin{pmatrix}0 & -1\\ 1 & z_j\end{pmatrix} L_j \begin{pmatrix}0 & -1\\ 1 & z_j\end{pmatrix}^{-1} = L_{j+1}.$$

We denote by $A_{\frac{j}{2k}}$ the matrix

$$A_{\frac{j}{2k}} = \begin{pmatrix}\alpha & \beta\\ \gamma & \delta\end{pmatrix}\begin{pmatrix}0 & -1\\ 1 & z_1\end{pmatrix}\cdots\begin{pmatrix}0 & -1\\ 1 & z_j\end{pmatrix}.$$

Then it is obvious that $ALA^{-1}=L'$ implies

$$A_{\frac{j}{2k}} L A_{\frac{j}{2k}}^{-1} = L',$$

Let $L=L_1$, $L' \in$ IV (which does not restrict the generality of the proof), and let the matrices A_i, A_{-i} satisfy Lemma VI.2.2. We consider the matrix

$$A_{i+\frac{-1}{2k}} = A_i\begin{pmatrix}z_i & 1\\ -1 & 0\end{pmatrix} = \begin{pmatrix}\alpha_i z_i-\beta_i & \alpha_i\\ \gamma_i z_i-\delta_i & \gamma_i\end{pmatrix}.$$

It is clear that for sufficiently large i (by the calculation (IV.2.3)) we have

$$(\alpha_i z_i-\beta_i)\beta_i<0, \quad (\gamma_i z_i-\delta_i)\delta_i<0, \quad L_2 \in \text{III}.$$

The matrix $-A_{i-\frac{2}{2k}}$ also satisfies Lemma VI.2.2, since $z_2<0$. We remark that in the matrix A_i

$$\alpha_i\gamma_i>0, \quad \beta_i\delta_i>0$$

and in the matrix A_{-i}

$$\alpha_{-i}\gamma_{-i}<0, \quad \beta_{-i}\delta_{-i}<0.$$

Let the matrix $A_{i-\frac{2n}{2k}}$ satisfy (VI.2.5) but not the matrix $A_{i-\frac{2n+1}{2k}}$.

We demonstrate that if the matrix $A_{i-\frac{2n+1}{2k}}$ does not fulfill Lemma VI.2.3, then the matrix $-A_{i-\frac{2(n+1)}{2k}}$ does satisfy this lemma. Denote by x, y, x', y' the components of the matrix $A_{i-\frac{2n+1}{2k}}$.

In case the matrix $A_{i-\frac{2n+1}{2k}}$ does not satisfy Lemma VI.2.3, we have

$$x<0,\ y>0,\ x'<0,\ y'>0$$

(assuming, finally, that the components of $A_{i-\frac{2n}{2k}}$ are all positive); it follows from this that $L', L_{i-\frac{2n}{2k}} \in \mathrm{IV}$, $y'>0$, and the matrix $A_{i-\frac{2n}{2k}}$ satisfies (VI.2.5). Therefore,

$$x<0,\ y>0,\ x'<0,\ y'>0,$$

$$A_{i-\frac{2(n+1)}{2k}} = \begin{pmatrix} x & y \\ x' & y' \end{pmatrix} \begin{pmatrix} z & 1 \\ -1 & 0 \end{pmatrix} \qquad (z<0).$$

Since $zx-y<0$ (this follows from $L' \in \mathrm{IV}$, $L_{i-\frac{2n+1}{2k}} \in \mathrm{III}$, $x<0$), and since the matrix $A_{i-\frac{2(n+1)}{2k}}$ does not satisfy (VI.2.5), we have $zx'-y'>0$; Lemma VI.3.3 is entirely proved.

We shall call the matrix A suitable if it satisfies Lemma VI.2.3. It is obvious that the third component of a suitable matrix possesses a sign opposite to the remaining components.

We note that if $L' \in \mathrm{III}$, then Lemma VI.2.3 remains true, but the second component of a suitable matrix will have in this case a different sign from the remaining components.

Let $L' \in \mathrm{II}$, $\det A = N > 0$, $x^2 \equiv D \pmod{N}$, and A suitable. Then the following lemma is valid.

Lemma VI.2.4: *For any matrix $L' \in \mathrm{II}$ there exists a matrix $L \in \mathrm{II}$ such that $ALA^{-1} = L'$ with A being suitable; in addition to this, if*

$$\frac{\sqrt{D}}{|a|},\ \frac{\sqrt{D}}{|c|},\ \frac{\sqrt{D}}{|a'|},\ \frac{\sqrt{D}}{|c'|} < k,$$

then

$$|\alpha|,\ |\beta|,\ |\gamma|,\ |\delta| < 2k\sqrt{N}.$$

The lemma is easily proved if we use (VI.2.3), the relations

$$\alpha\delta \leqq N, \quad -\beta\gamma \leqq N,$$

and the condition that A is a suitable matrix.

Let $\det A = N > 0$, $ALA^{-1} = L'$ and let the matrix A be suitable. We set

$$L_{2k} = L_0, L_1, \ldots, L_{2k},$$
$$L'_{2k} = L'_0, L'_1, \ldots, L'_{2k};$$

then we obtain

$$(***)\qquad \begin{aligned} A L_0 A^{-1} &= L'_0 \\ A_{-\frac{1}{2k}} L_1 A^{-1}_{-\frac{1}{2k}} &= L'_0 \\ \vdots \qquad & \quad \vdots \\ A_{-\frac{i}{2k}} L_i A^{-1}_{-\frac{i}{2k}} &= L'_0 . \end{aligned}$$

Remark. We discern easily from (VI.2.5) and (VI.2.6) that if the matrix A is suitable, but the matrix $A_{-\frac{i}{2k}}$ not suitable, then the matrix $A_{-\frac{j}{2k}}$ is not suitable for all $j \geqq i$.

Let the matrices $A, A_{-\frac{1}{2k}}, \ldots, A_{-\frac{c}{2k}}$ be suitable and the matrix $A_{-\frac{c+1}{2k}}$ not suitable ($c < i$). Since

$$A_{-\frac{n+1}{2k}} = A_{-\frac{n}{2k}} \begin{pmatrix} z_{-\frac{n}{2k}} & 1 \\ -1 & 0 \end{pmatrix}$$

and since each component of a suitable matrix is bounded by the determinant of the matrix, c must be of the order $0(\ln N)$.

Handling each of the matrices L_j' like L_0' in (***) and taking into account the remark above, we also see that for each such matrix there exist no more than $0(\log N)$ matrices L_i for which the matrix $A(AL_iA^{-1} = L_j')$ is suitable. Thus we obtain the following theorem.

Theorem VI.2.2 *(theorem on cycles): Let the matrix A with* $\det A = N > 0$ *be such that $ALA^{-1} = L'$; then*

$$\frac{2k}{2k'} < c_1 \log(1+N),$$

where c_1 is an absolute constant.

Suppose that $L, L' \in$ II. Obviously there exists a class of forms such that

$$b_0 + L = P_0 Q_0, \quad b_0 - L' = Q_0 P_0,$$

with representative

$$(\det P_0, b_0, \det Q_0), \quad b_0^2 - \det P_0 \cdot \det Q_0 = D.$$

Also, it can be assumed that for a given matrix L', the matrix L is such that

$$0 < b_0 < \sqrt{D}, \quad \sqrt{D} - b_0 < \det Q_0 < \sqrt{D} + b_0,$$

$$P_0 = \begin{pmatrix} n_0' & l_0' \\ m_0' & k_0' \end{pmatrix}, \quad Q_0 = \begin{pmatrix} k_0 & l_0 \\ m_0 & n_0 \end{pmatrix},$$

$$|l_0| > |k_0|, \quad |n_0| > |m_0|,$$

$$|l_0'| > |k_0'|, \quad |n_0'| > |m_0'|.$$

Without loss of generality we may consider the case where $L' \in$ IV, $L \in$ III. We set $-L' = L''$. Since $a < 0$, $c > 0$, $a'' < 0$, $c'' > 0$, $b'' < 0$, we have

$$\text{(VI.2.7)} \qquad \begin{cases} \dfrac{|l_0|}{|k_0|} < |\omega_1|, & \dfrac{l_0}{k_0}\omega_1 > 0, \\[2mm] \dfrac{|n_0|}{|m_0|} > |\omega_1|, & \dfrac{n_0}{m_0}\omega_1 > 0, \\[2mm] \dfrac{|l_0'|}{|k_0'|} > |\omega_1|, & \dfrac{l_0'}{k_0'}\omega_1 < 0, \\[2mm] \dfrac{|n_0'|}{|m_0'|} < |\omega_1|, & \dfrac{n_0'}{m_0'}\omega_1 < 0, \end{cases}$$

$$\text{where } \omega_1 = \frac{-b - \sqrt{D}}{c}.$$

Further, we have

$$\text{(VI.2.8)}\qquad \begin{cases} \dfrac{|k_0|}{|m_0|} > \omega_1', & \dfrac{k_0}{m_0}\,\omega_1' < 0\,,\\ \dfrac{|l_0|}{|n_0|} < \omega_1', & \dfrac{l_0}{n_0}\,\omega_1' < 0\,,\\ \dfrac{|k_0'|}{|m_0'|} < \omega_1', & \dfrac{k_0'}{m_0'}\,\omega_1' > 0\,,\\ \dfrac{|l_0'|}{|n_0'|} > \omega_1', & \dfrac{l_0'}{n_0'}\,\omega_1' > 0\,, \end{cases}$$

$$\text{where } \omega_1' = \frac{-b' - \sqrt{D}}{c'}\,.$$

Let $k_0 > 0$ (which is always easy to attain); then it follows from (VI.2.7) and (VI.2.8) that

$$k_0 > 0,\ l_0 < 0,\ m_0 < 0,\ n_0 > 0\,.$$

From the equation

$$\text{(VI.2.9)}\qquad \begin{pmatrix} b_0 + b & -a \\ c & b_0 - b \end{pmatrix} = P_0 Q_0$$

we obtain

$$n_0' > 0,\ l_0' > 0,\ m_0' > 0,\ k_0' > 0\,.$$

According to (VI.2.9), we put

$$\frac{n_0'}{n_0} \cdot \det Q_0 = b_0 - b' + \frac{l_0}{n_0}\,c'\,.$$

Since $l_0/n_0 > \omega_1'$,

$$\frac{n_0'}{n_0} < \frac{b_0 + \sqrt{D}}{\det Q_0} = -\omega_1^{(0)}\,.$$

Analogously we find the relations

$$\text{(VI.2.10)}\qquad \frac{k_0'}{k_0} < -\omega_1^{(0)},\ \frac{l_0'}{l_0} < -\omega_1^{(0)},\ \frac{m_0'}{m_0} < -\omega_1^{(0)},\ \frac{n_0'}{n_0} < -\omega_1^{(0)}\,.$$

We remark that the right neighbor of the form $(\det P_0, b_0, \det Q_0)$ turns out to be the form $(\det Q_0, b_1, \det Q_1)$, $\det Q_1 < 0$.

We have

$$\bar{P}_1 = -Q_0\,,$$
$$Q_1 = P_0 + Z_0 Q_0$$

where $\bar{P}P = \det P$, $|Z_0| = [-\omega_1^{(0)}]$, $Z_0 < 0$, and this means

$$b_1 + L = P_1 Q_1 = -\bar{Q}_0(\bar{P}_0 + Z_0 Q_0) = -b_0 - Z_0 \det Q_0 + L\,.$$

We clarify the structure of P_1 and Q_1. Obviously,

$$P_1 \to \begin{pmatrix} - & - \\ - & - \end{pmatrix}.$$

Let the components of Q_0 be sufficiently large in absolute value (this is always realizable). Then, using (VI.2.10), we shall have

$$Q_1=\begin{pmatrix} k_0'+Z_0k_0 & -l_0'+Z_0l_0 \\ -m_0'+Z_0m_0 & n_0'+Z_0n_0 \end{pmatrix}\to\begin{pmatrix} + & - \\ - & + \end{pmatrix}.$$

Therefore, we get

$$P_1\to\begin{pmatrix} - & - \\ - & - \end{pmatrix},\quad Q_1\to\begin{pmatrix} + & - \\ - & + \end{pmatrix}$$

and for P_1, Q_1, $\det Q_1<0$ a formula analogous to the formula (VI.2.10) will be valid:

$$\text{(VI.2.11)}\qquad \frac{|k_1'|}{|k_1|}>\omega_1^{(1)},\ \frac{|l_1'|}{|l_1|}<\omega_1^{(1)},\ \frac{|m_1'|}{|m_1|}<\omega_1^{(1)},\ \frac{|n_1'|}{|n_1|}<\omega_1^{(1)},$$

where $\omega_1^{(1)}=\dfrac{-b_1-\sqrt{D}}{\det Q_1}$.

We consider now the general scheme of transition from (P_i, b_i, Q_i) to $(P_{i+j}, b_{i+j}, Q_{i+j})$. We begin with (P_i, b_i, Q_i) and $\det Q_i>0$. We have

$$\begin{aligned}
&P_i\to\begin{pmatrix} + & + \\ + & + \end{pmatrix},\quad Q_i\to\begin{pmatrix} + & - \\ - & + \end{pmatrix},\quad \det Q_i>0,\ \omega_1^{(i)}<0\\
\text{(VI.2.12)}\quad &P_{i+1}\to\begin{pmatrix} - & - \\ - & - \end{pmatrix},\quad Q_{i+1}\to\begin{pmatrix} + & - \\ - & + \end{pmatrix},\quad \det Q_{i+1}<0,\ \omega_1^{(i+1)}>0\\
&P_{i+2}\to\begin{pmatrix} + & + \\ + & + \end{pmatrix},\quad Q_{i+2}\to\begin{pmatrix} + & - \\ - & + \end{pmatrix},\quad \det Q_{i+2}>0,\ \omega_1^{(i+2)}<0
\end{aligned}$$

etc. This transition process can not be infinite without a sign violation, since the components of the matrices decrease.

From (VI.2.10) and (VI.2.11) it follows that such a violation can happen either in the fourth sign of Q_j, if $\det Q_j<0$, or in the third sign of Q_j, if $\det Q_j>0$. Let P_0Q_0 be the last pair in (VI.2.12) in which no sign violation occurs. Then, applying (VI.2.7)—(VI.2.12), we obtain the theorem.

Lemma VI.2.5: *Let $L, L'\in \mathrm{II}$ and* $\dfrac{\sqrt{D}}{|a|}, \dfrac{\sqrt{D}}{|c|}, \dfrac{\sqrt{D}}{|a'|}, \dfrac{\sqrt{D}}{|c'|}<k$; *then there exists a properly primitive reduced quadratic form*

$$(\det P_1, b_1, \det Q_1)\in \mathrm{II},\ b_1^2-\det P_1\cdot\det Q_1=D,$$

for which

$$b_1+L=P_1Q_1,\ b_1-L'=Q_1P_1$$

and in addition

$$P_1=0(2kD^{\frac{1}{4}}),\ Q_1=0(2kD^{\frac{1}{2}}|\det P_1|^{-\frac{1}{2}}),$$

or

$$P_1=0(2kD^{\frac{1}{2}}|\det Q_1|^{-\frac{1}{2}}),\ Q_1=0(2kD^{\frac{1}{4}}),$$

where the expression $P_1, Q_1=0(M)$ signifies that all components of the given matrix are bounded in absolute value by the number M.

Lemma VI.2.6: *For any M and arbitrarily small $\eta > 0$ the number of vector matrices $L \in \mathrm{II}$ for which $\frac{\sqrt{D}}{|c|}$ or $\frac{\sqrt{D}}{|a|} > M$ is of the order*

$$0(M^{-1} D^{\frac{1}{2}+\eta}).$$

Proof: From the reduction condition it follows that if $(a, b, c) \in \mathrm{II}$ and $(a, b', c') \in \mathrm{II}$, then $b \not\equiv b' \pmod{a}$. Therefore the number of such forms with given a has the order

$$0(D^{\frac{\eta}{2}} \sqrt{(D, a)});$$

this implies that the number which we are interested in possesses the estimate

$$2M^{-1} \cdot D^{\frac{1}{2}} \sum_{r|D,\, r \leqq M^{-1} D^{\frac{1}{2}}} r^{-\frac{1}{2}} 0(D^{\frac{\eta}{2}}),$$

and since

$$\sum_{r|D} r^{-\frac{1}{2}} \leqq \sum_{r|D} 1 = 0(D^{\frac{\eta}{2}}),$$

the required estimate is valid.

In conclusion of this point we make the following remarks about rotations.

1. Let $L = L_1, L_2, \ldots, L_{2k}$ and $L' = L'_1, L'_2, \ldots, L'_{2k'}$ be the matrices of Region II, numbered in the order designated earlier. We denote suitable matrices by Latin capitals and two subscripts, and we set

$$\text{(VI.2.13)} \quad \begin{aligned} A_{ij} L_i A_{ij}^{-1} &= A_{i+1,j} L_{i+1} A_{i+1,j}^{-1} \\ &= \cdots = A_{i+n,j} L_{i+n} A_{i+n,j}^{-1} = L'_j \qquad (n = 0, 1, 2, \ldots). \end{aligned}$$

Then it is clear from the preceding that a rotation

$$A_{i+n,j+n'} L_{i+n} A_{i+n,j+n'}^{-1} = L'_{j+n'} \quad \text{for} \quad n' > 0,\ n > 0$$

does not exist.

In the situation above we shall say that the matrix L_{i+n} belongs to the matrix L'_j according to the number $\det A_{ij}$ (for brevity we shall sometimes omit "matrix" and "number") and we shall write symbolically for this:

$$(L'_j;\ L_{i+n}) \det A.$$

If

$$A_{ij} L_i A_{ij}^{-1} = A_{i,j+1} L_i A_{i,j+1}^{-1} = \cdots = A_{i,j+n'} L_i A_{i,j+n'}^{-1} = L'_{j+n'} \quad (n' = 0, 1, 2, \ldots),$$

then by the same considerations a rotation

$$A_{i+n,j+n'} L_{i+n} A_{i+n,j+n'}^{-1} = L'_{j+n'} \quad \text{for} \quad n > 0,\ n' > 0$$

does not exist. This is written symbolically as

$$(L_i;\ L'_{j+n'}) \det A.$$

If we wish to express the number of matrices $L'_{j+n'}$ or L_{i+n} belonging to the given matrix L_i or L'_j respectively, then we write $(i;\ j+n')$ or $(j;\ i+\mathrm{n})$; this symbol expresses also that the matrices $L'_{j+n'+1}$ and

L'_{j-1} do not belong to L_i and that the matrices L_{i+n} and L_{i-1} do not belong to L'_j.

Suppose we have $(1'; 1+n)$; then it follows immediately from the preceding that either

$$(L'_2;\ L_{n_1+2})$$

or

$$(L_{n_1+2};\ L'_2)$$

is fulfilled. In this case we shall say that the suitable matrix performs a right revolution.

2. Let the matrices $L = L_1, L_2, \ldots, L_{2k}$ and $L' = -L'_1, -L'_2, \ldots, -L'_{2k'}$ be such that the matrices $L = L_1, L_2, \ldots, L_{2k}$ and $-L' = L'_1, L'_2, \ldots, L'_{2k'}$ lie on the Region II, numbered in the agreed manner. We consider the rotation $ALA^{-1} = -L'$ and introduce notation analogous to the notation for the rotation $ALA^{-1} = +L'$. We assume that $(1'; 1+n_1)$ is satisfied; then on the basis of the preceding either

$$(-L'_2;\ L_{2k}) = (-L'_2;\ L_{1-1})$$

or

$$(L_{n_1+2};\ -L'_{2k'}) = (L_{n_1+2};\ -L'_{1-1})$$

is satisfied. In this case we shall say that the suitable matrix performs a left revolution.

§ 3. Asymptoticity of Hyperbolic Rotations

Let $p \geqq 3$ be a fixed prime number such that $(D/p) = +1$. Then the congruence $l^2 \equiv D \pmod{p^s}$ is solvable for any $s > 0$. We choose s so that

$$D^{\frac{1}{2}+\eta_2} < p^s \leqq p\, D^{\frac{1}{2}+\eta_2},$$

where $\eta_2 > 0$ is an arbitrary small number. Then

$$l^2 - D = p^s x,$$

where l is chosen such that $(x, p) = 1$. We write down the $T(D)$ equations

$$l + L'_i = \Pi_i X_i, \quad i = 1, 2, \ldots, T(D)$$
$$(\det \Pi_i = p^s, \ \det X_i = x),$$

and we shall assume that all matrices Π_i are suitable; in this case we have

(VI.3.1) $$l + L_j = X_j \Pi_j \quad (j = 1, 2, \ldots, T(D)), \quad L_j \in \mathrm{II}.$$

Remark. Among the equations (VI.3.1) different indices may occur for which the corresponding L_j are equal. But if we perform the rotation on L_j with the help of its corresponding matrix Π_j, then we obtain matrices $L'_i \in \mathrm{II}$ which are all different; then in the equations (VI.3.1) the number of identical L_j does not exceed $0(\log D)$.

Let $0 < \eta_1 < \eta_2$; the exact value of η_1 in terms of η_2 $(\eta_1 = k_1 \eta_2)$ we shall indicate in the following. We consider any T_1 of the equations (VI.3.1), where we set $T_1 > \mathrm{D}^{\frac{1}{2}-\eta_1}$, and we number them consecutively from 1 to T_1. We have

(VI.3.2) $$l + L_j = X_j \Pi_j \qquad (j = 1, 2, \ldots, T_1).$$

We show that the number of different Π_j in (VI.3.2) can not be less than $D^{\frac{1}{2}-\eta_3}$ ($\eta_3 = k_2 \eta_2$; the exact value of η_3 will be chosen later). We set $M = D^{\eta_1 + 2\eta}$ in Lemma VI.2.6. The number of matrices L_i' for which $\frac{\sqrt{D}}{|c'|} > M$ or $\frac{\sqrt{D}}{|a'|} > M$ will be of the order $0(D^{\frac{1}{2}-\eta_1-\eta})$. For the remaining matrices L_i' we obviously have

$$\frac{\sqrt{D}}{|c'|}, \ \frac{\sqrt{D}}{|a'|} < M.$$

Hence it follows that for sufficiently large D there will not be less than $T_2 \geqq \frac{1}{2} T_1 \geqq \frac{1}{2} D^{\frac{1}{2}-\eta_1}$ equations among the T_1 equations (VI.3.2) such that for the corresponding matrices L_i' the condition

$$\frac{\sqrt{D}}{|c'|}, \ \frac{\sqrt{D}}{|a'|} < M \leqq D^{\eta_1 + \eta'}$$

is fulfilled, where $\eta' > 0$ is arbitrarily small.

Let us now take instead of equations (VI.3.2) the above T_2 equations

(VI.3.3) $$l + L_j = X_j \Pi_j \qquad (j = 1, 2, \ldots, T_2)$$

for which

(VI.3.4) $$\frac{\sqrt{D}}{|c'|}, \ \frac{\sqrt{D}}{|a'|} < D^{\eta_1 + \eta'}.$$

Let $P_1, P_2, \ldots, P_w$ be all of the different matrices Π_j occurring in the equations (VI.3.3). Dividing the equations (VI.3.3) into w systems $G_1, G_2, \ldots, G_w$, each with one such P_j, we get

$$\begin{aligned} l + L_j &= Q_j P_1 \quad (j = 1, 2, \ldots, a_1) \\ l + L_j &= Q_j P_2 \quad (j = a_1 + 1, \ldots, a_1 + a_2) \\ &\vdots \\ l + L_j &= Q_j P_w \quad (j = a_1 + \cdots + a_{w-1} + 1, \ldots, a_1 + \cdots + a_w) \\ T_2 &= a_1 + \cdots + a_w . \end{aligned}$$

We show that the number of distinct Π_j is not less than $D^{\frac{1}{2}-\eta_3}$. We assume the contrary, that is, $w < D^{\frac{1}{2}-\eta_3}$, and we reduce this assumption to a contradiction.

As we established above, in the equations (VI.3.3) the number of identical L_j (for distinct indices) does not exceed $0(\log D)$. Further, the number of L_j not fulfilling the condition

$$\frac{\sqrt{D}}{|c|}, \ \frac{\sqrt{D}}{|a|} < D^{\eta_1 - \eta'}$$

(with repetitions) is of the order

$$0(D^{\frac{1}{2}-\eta_1-\eta'} \log D),$$

or, for sufficiently large D

$$0(D^{\frac{1}{2}-\eta_1-\frac{\eta'}{2}}).$$

Discarding all equations $l+L_j=Q_j\cdot P_j$ with such L_j, we obtain $T_3\geqq\frac{1}{2}T_2\geqq\frac{1}{4}D^{\frac{1}{2}-\eta_1}$ equations, in which each L_j satisfies the condition

(VI.3.5) $$\frac{\sqrt{D}}{|a|},\ \frac{\sqrt{D}}{|c|}<D^{\eta_1-\eta'};$$

instead of T_3 we shall simply write $T_2\geqq\frac{1}{4}D^{\frac{1}{2}-\eta_1}$. Together with each equation

$$l+L_j=Q_jP_k$$

we now have the equation

$$l+L'_j=P_kQ_j,$$

where $L'_j\in$ II and satisfies (VI.3.5). Taking conjugates, we obtain the equation

$$l-L'_j=\bar{Q}_j\bar{P}_k.$$

Therefore, to each system

$$G:l+L_j=Q_jP_k\quad (j=a_1+\cdots+a_{k-1}+1,\dots,a_1+\cdots+a_k)$$

corresponds the system

$$\bar{G}:l-L'_j=\bar{Q}_j\bar{P}_k\quad (j=a_1+\cdots+a_{k-1}+1,\dots,a_1+\cdots+a_k).$$

According to Lemma VI.2.5, there exists a form $(a,b,c)\in$ II, with corresponding matrices A, B, $\det A=a$, $\det B=c$, for which

$$AL_jA^{-1}=-L'_j,\quad B^{-1}L_jB=-L'_j,\quad b+L_j=AB,\quad b-L'_j=BA,$$

and in addition

(VI.3.6) $$A=0(M\cdot D^{\frac{1}{4}}),\ B=0(M\cdot D^{\frac{1}{2}}|a|^{-\frac{1}{2}})$$

or

(VI.3.7) $$A=0(MD^{\frac{1}{2}}|c|^{-\frac{1}{2}}),\ B=0(M\cdot D^{\frac{1}{4}}).$$

Consequently,

$$G_k:l+L_j=Q_jP_k;\ \bar{G}_k:l-L'_j=\bar{Q}_j\bar{P}_k;$$
$$b+L_j=AB,\ b-L'_j=BA,$$

and therefore

$$P_kA\equiv P_k\bar{B}\equiv 0\,(\mathrm{mod}\,\bar{P}_k\ \text{right}),$$

$$\mathrm{Sp}(P_kA)\equiv 0\,(\mathrm{mod}\,p^s),\ \mathrm{Sp}(P_k\bar{B})=\mathrm{Sp}(B\bar{P}_k)\equiv 0\,(\mathrm{mod}\,p^s).$$

We shall assume that the condition (VI.3.6) is satisfied (otherwise we place indices on the hermition $H:H_1=P_k\mathrm{A}$, $H_2=B\bar{P}_k$). Then

(VI.3.8) $$\mathrm{Sp}(H_1)\equiv\mathrm{Sp}(H_2)\equiv 0\,(\mathrm{mod}\,p^s).$$

We have

$$H_1=g+g_1i_1+g_2i_2+g_3i_3\quad (2g_i\ \text{integral})$$
$$H_2=g'+g'_1i_1+g'_2i_2+g'_3i_3\quad (2g'_i\ \text{integral}),$$

where i_1, i_2, i_3 are unit hermitions, corresponding to the form $x_1^2 - x_2 x_3$ (see Chapter V). We estimate the absolute values of the components according to the choice of (a, b, c). Taking (VI.3.6), (VI.3.5) and (VI.3.4) into account,

$$H_1 = 0(p^{\frac{s}{2}} D^{\frac{1}{4}} M^2) = 0(p^{\frac{s}{2}} D^{\frac{1}{4}} D^{2\eta_1 + 2\eta'}),$$
$$H_2 = 0(p^{\frac{s}{2}} D^{\frac{1}{2}} |a|^{-\frac{1}{2}} D^{2\eta_1 + 2\eta'}).$$

Obviously, in order for the equation $\mathrm{Sp}(H_1) = 0$ to be valid, it is sufficient that the condition

$$p^{\frac{s}{2}} D^{\frac{1}{4}} D^{2(\eta_1 + \eta')} < p^s$$

be fulfilled. Since $p^s > D$,

$$D^{\frac{1}{4}} D^{2(\eta_1 + \eta')} \leqq D^{\frac{1}{4} + \frac{\eta_2}{2}},$$

i.e.,

$$4(\eta_1 + \eta') \leqq \eta_2 .$$

We set $\eta_2 = 4.6\eta_1$, $\eta' \leqq 0.1\eta_1$; then we obtain

$$4(\eta_1 + 0.1\eta_1) < 4.6\eta_1 .$$

Then the estimates

$$\text{(VI.3.9)} \qquad \begin{aligned} &H_1 = 0(p^{\frac{s}{2}} D^{\frac{1}{4} + 2.2\eta_1}), && \mathrm{Sp}(H_1) = g = 0 \\ &H_2 = 0(p^{\frac{s}{2}} D^{\frac{1}{2} + 2.2\eta_1} |a|^{-\frac{1}{2}}), && \mathrm{Sp}(H_2) = g' = \frac{dp^s}{2} \end{aligned}$$

are valid.

We go over to the rotations. Here we have

$$H_1 = g_1 i_1 + g_2 i_2 + g_3 i_3 \quad (2g_i \text{ integral}),$$
$$H_2 = \frac{dp^s}{2} + g_1' i_1 + g_2' i_2 + g_3' i_3 \quad (2g_i' \text{ integral}).$$

Consider the quadratic forms

$$\mathrm{Norm}(H_1 x + \bar{H}_2 y) = p^s(ax^2 + 2bxy + cy^2),$$
$$\mathrm{Norm}(H_1 x + H_2 y) = \left(\frac{dp^s}{2} y\right)^2 + (g_1 x + g_1' y)^2$$
$$- (g_2 x + g_2' y)^2 - (g_3 x + g_3' y)^2 .$$

It is clear that

$$(g_1 x + g_1' y)^2 - (g_2 x + g_2' y)^2 - (g_3 x + g_3' y)^2 =$$
$$= p^s\left(ax^2 + 2bxy + \left(c - \frac{d^2 p^s}{4}\right) y^2\right),$$

that is,

$$\text{(VI.3.10)} \qquad \begin{aligned} &(2g_1 x + 2g_1' y)^2 - (2g_2 x + 2g_2' y)^2 - (2g_3 x + 2g_3' y)^2 = \\ &\qquad = p^s(4ax^2 + 2 \cdot 4bxy + (4c - d^2 p^s) y^2) . \end{aligned}$$

We compute the number of initial conjugate pairs $(L_i; (-L_j'))$ corresponding to the given numbers d, g_1, g_2, g_3, g_1', g_2', g_3', if it is taken into account that L_i and L_i' correspond to the condition (VI.3.5). We remark that one of the hermitions $H(H_1$ or $H_2)$ is necessarily primitive. Fixing the above mentioned numbers determines H_1 and H_2.

Without loss of generality we assume that H_1 is primitive; if P_k^0, A_1^0 are solutions of the equation $P_k^0 A_1^0 = H_1$, then all other solutions will be of the form

$$P_k^0 \bar{\varepsilon} = P_k', \quad \varepsilon A_1^0 = A'.$$

Further, $H_2 = B^1 \bar{P}_k'$; that is, $B' = B^0 \bar{\varepsilon}$. Thus,

$$A^0 B^0 = b + L, \quad A' B' = b + \varepsilon L \bar{\varepsilon} = b + L_1,$$

hence we obtain

$$l + L = Q_i^0 P_k^0, \quad l + L_1 = \varepsilon Q_i^0 P_k^0 \bar{\varepsilon}.$$

But this contradicts the choice of (VI.3.1) for $\varepsilon \neq \pm 1$. Consequently, the given numbers determine no more than two conjugate pairs $(L_i; (-L'))$.

We return to the representation (VI.3.10). It is obvious that the number p^s divides the g.c.d. of the three determinants

(VI.3.11) $$\begin{vmatrix} 2g_2 & 2g_2' \\ 2g_3 & 2g_3' \end{vmatrix}, \quad \begin{vmatrix} 2g_3' & 2g_3 \\ 2g_1' & 2g_1 \end{vmatrix}, \quad \begin{vmatrix} 2g_1 & 2g_1' \\ 2g_2 & 2g_2' \end{vmatrix}.$$

We compute the determinant of the quadratic form (VI.3.10). We have

$$\Delta = p^{2s}(16b^2 - 4a(4c - d^2 p^s)) = p^{2s}(16D + 4ad^2 p^s).$$

Hence it follows that if $\Delta < 0$, then $a < 0$ and the form (VI.3.10) is negative. If $\Delta > 0$, then the form (VI.3.10) is non-negative.

Thus, in both cases there exist representations of the above mentioned quadratic form as a ternary form $x^2 - y^2 - z^2$. We estimate Δ. Since according to (VI.3.9)

$$p^s d = 0(p^{\frac{s}{2}} D^{\frac{1}{2} + 2.2\eta_1} |a|^{-\frac{1}{2}}),$$

we have

(VI.3.12) $$\Delta = 0(p^{3s} D^{1 + 4.4\eta_1}).$$

There may be infinitely many representations of the type (VI.3.10) if no restrictions are placed on the components. If H_1 and H_2 are hermitions, then the equation $H_1 H_2 = \xi + M$, where M is a vector and ξ an integer, gives a representation of the form

$$\mathrm{Norm}(H_1) x^2 + 2\xi xy + \mathrm{Norm}(H_2) y^2$$

by the form $x^2 - y^2 - z^2$ (here the form may be either negative or non-negative).

If $\mathrm{Norm}(\varepsilon) = +1$, then

$$\varepsilon H_1 \bar{\varepsilon} \cdot \varepsilon H_2 \bar{\varepsilon} = \xi + \varepsilon M \bar{\varepsilon},$$

and we obtain analogous representations for any integral ε. The representations (VI.3.10) are classified according to restrictions imposed on the components. Those restrictions are expressed in the conditions (VI.3.9) and (VI.3.12).

Thus, we have a representation (VI.3.10) under the conditions

$$
\text{(VI.3.13)} \qquad \begin{cases} p^s d = 0(p^{\frac{s}{2}} |a|^{-\frac{1}{2}} D^{\frac{1}{2}+2.2\eta_1}), \\ 2g_1, 2g_2, 2g_3 = 0(p^{\frac{s}{2}} D^{\frac{1}{4}+2.2\eta_1}), \\ (2g_1, 2g_2, 2g_3) = 2p^{s_1} \text{ or } p^{s_1} \\ (4a, 4b, 4c - d^2 p^s) = e_1, \text{ i.e.,} \\ (4a, D) = e_2 \geqq \frac{1}{4} e_1 . \end{cases}
$$

The total number of integral forms for all values of e is of the order

$$
\text{(VI.3.14)} \qquad 0(D^{\xi})(e_2)^{\frac{1}{2}},
$$

where $\xi > 0$ is as small as desired (we do not present the derivation of this formula, since it is completely analogous to the derivation given in Chapter V; the only difference is that in Chapter V it was sufficient to write down the matrices (V.6.16) and (V.6.17) for $\kappa = 1$ and $\kappa = 2$; here we would have to consider them for all $\kappa | e$.

We estimate the number of primitive representations of an arbitrary form of the above type, for example Ψ_λ, by $x^2 - y^2 - z^2$. We have

$$
\text{(VI.3.15)} \qquad \begin{cases} \Psi_\lambda = (ax + by)^2 - (a'x + b'y)^2 - (a''x + b''y)^2 , \\ a i_1 + a' i_2 + a'' i_3 = \dfrac{2g_1}{\kappa} i_1 + \dfrac{2g_2}{\kappa} i_2 + \dfrac{2g_3}{\kappa} i_3 , \\ b i_1 + b' i_2 + b'' i_3 = \dfrac{2g_1' - \lambda 2d_1}{\mu} i_1 + \\ \qquad + \dfrac{2g_2' - \lambda 2g_2}{\mu} i_2 + \dfrac{2g_3' - \lambda 2g_3}{\mu} i_3 , \\ \mu = \dfrac{l_1}{\kappa} , \end{cases}
$$

or

$$
\text{(VI.3.16)} \qquad \begin{aligned} a i_1 + a' i_2 + a'' i_3 &= \frac{2H_1}{\kappa} , \\ b i_1 + b' i_2 + b'' i_3 &= \frac{M}{\mu} - \frac{\lambda}{\mu} 2H_1 , \end{aligned}
$$

where $M = 2H_2 - dp^s$. In addition, $H_1 H_2 = p^s(b + L_j')$, where $L_j' \in \text{II}$.

In order to obtain all primitive representations (VI.3.15) we shall have to find all systems of Gaussian numbers M, N characterizing these representations (see VENKOV [2], p. 139). We denote the determinant of $\Psi_\lambda = \Psi_\lambda(p, q, r)$ by $\Omega \cdot \lambda$; then M and N will satisfy the congruences

$$
-p \equiv N^2, \quad q \equiv MN, \quad -r \equiv M^2 \pmod{\Omega \cdot \lambda} .
$$

The number of pairs M, N modulo $\Omega \cdot \lambda$ will be of the order

$$
\text{(VI.3.17)} \qquad 0(D^{\xi_1})
$$

where ξ_1 is a constant as small as desired.

We consider any pair of numbers M, N and the representation (VI.3.15) belonging to $\pm M$, $\pm N$. In the worst case for the upper estimate, the ternary form g constructed from the numbers M, N (see VENKOV [2], p. 138) will be integral and equivalent to the form $f = x^2 - y^2 - z^2$. If $fs = g$, where

$$s = \begin{pmatrix} a & b & c \\ a' & b' & c' \\ a'' & b'' & c'' \end{pmatrix}$$

is a unimodular integral matrix, then all primitive representations (VI.3.15), belonging to $\pm M$, $\pm N$, will be given by the first two columns of this matrix. It is known that if $f \cdot s = g$ and $f \cdot s_1 = g$, then $s \cdot s_1^{-1}$ is an automorphism of f. Denoting by u, u', ... the automorphisms of f, we find that

$$ss_1^{-1} = u, \quad s_1 = u^{-1}s = u'\begin{pmatrix} a & b & c \\ a' & b' & c' \\ a'' & b'' & c'' \end{pmatrix}.$$

Therefore, the representation Ψ_λ is equivalent to a representation obtained by letting an automorphism operate on the form $x^2 - y^2 - z^2$, setting in it

(VI.3.18) $$X = ax + by, \quad Y = a'x + b'y, \quad Z = a''x + b''y.$$

Suppose

$$H_1' = ai_1 + a'i_2 + a''i_3;$$
$$M' = bi_1 + b'i_2 + b''i_3;$$

then taking (VI.3.18) into account, we get

$$X^2 - Y^2 - Z^2 = \mathrm{Norm}(Xi_1 + Yi_2 + Zi_3) = \mathrm{Norm}(H_1'x + M_y'').$$

We show that if the form $X^2 - Y^2 - Z^2$ is transformed by an automorphism, setting

$$X = ax + by, \quad Y = a'x + b'y, \quad Z = a''x + b''y,$$

then this will be equivalent to a substitution

$$H_1' \to \varepsilon H_1' \varepsilon^{-1}, \quad M' \to \varepsilon M' \varepsilon^{-1}, \quad \mathrm{Norm}(\varepsilon) = \pm 1.$$

The general form of integral automorphisms of the form $X^2 - Y^2 - Z^2$ with determinant $+1$ is given by the formula

$$X'i_1 + Y'i_2 + Z'i_3 = \varepsilon(Xi_1 + Yi_2 + Zi_3)\varepsilon^{-1},$$

where $\mathrm{Norm}(\varepsilon) = \pm 1$ and ε is an integral hermition.

By virtue of the statements above, we can obtain from the representation of Ψ_λ by f, using a substitution s, all other representations corresponding to the product

$$\varepsilon(H_1'X + M'Y)\varepsilon^{-1},$$

or, in expanded form

$$\varepsilon H_1' \varepsilon^{-1} X + \varepsilon M' \varepsilon^{-1} Y.$$

Here, H_1' is replaced by $\varepsilon H_1' \varepsilon^{-1}$. Turning to (VI.3.16), we see that H_1 is replaced by $\varepsilon H_1 \varepsilon^{-1}$ and

$$H_2 = \frac{\mu}{2} M' + 2\lambda H_1 + dp^s,$$

so that H_2 will go over in $\varepsilon H_2 \varepsilon^{-1}$. In view of this the product

$$H_1 H_2 = p^s(b + L_j')$$

will be replaced by the expression

$$\varepsilon H_1 \varepsilon^{-1} \varepsilon H_2 \varepsilon^{-1} = p^s(b + \varepsilon L_j' \varepsilon^{-1}). \tag{VI.3.19}$$

We estimate the number of units ε in (VI.3.19) for which $\varepsilon L_j' \varepsilon^{-1} \in \mathrm{II}$. We have

$$\begin{aligned} &H_1' \to \varepsilon H_1' \varepsilon^{-1}, \quad H_2' \to \varepsilon H_2' \varepsilon^{-1}, \\ &\varepsilon H_1' \varepsilon^{-1} = \varepsilon P_k \varepsilon_1^{-1} \varepsilon_1 A \varepsilon^{-1}, \\ &\varepsilon H_2' \varepsilon^{-1} = \varepsilon B \varepsilon_1^{-1} \varepsilon_1 P_k \varepsilon^{-1}. \end{aligned}$$

Here $\varepsilon P_k \varepsilon^{-1}$, $\varepsilon_1 A \varepsilon^{-1}$, and $\varepsilon B \varepsilon_1^{-1}$ satisfy the condition (VI.3.7) and the form (a, b, c) with the matrices A, B fulfills the condition in Lemma VI.2.4. In addition, on $\varepsilon P_k \varepsilon_1^{-1}$ the units ε and ε_1^{-1} must induce a right revolution; concerning $\varepsilon_1 A \varepsilon^{-1}$ and $\varepsilon B \varepsilon_1^{-1}$, in so far as $BLB^{-1} = A^{-1}LA = -L'$, the units induce left rotations. Hence it follows that the number of equations of type (VI.3.19) for given M, N does not exceed a constant times $\log D$.

From the above statements and by virtue of (VI.3.14), (VI.3.17) we deduce that for given d the number of representations (VI.3.10) has the estimate

$$0(D^{\xi})(e_2)^{\frac{1}{2}}$$

where ξ is arbitrarily small.

Therefore, taking $\xi = 0.1\eta_1$, we get that for the given form $\varphi = (a, b, c)$ and variable d, the number of equations (VI.3.10) has the order

$$0(p^{-\frac{s}{2}} D^{\frac{1}{2} + 2.3\eta_1} |a|^{-\frac{1}{2}})(e_2)^{\frac{1}{2}} \qquad (d > 0). \tag{VI.3.20}$$

According to § 2, the number of conjugate pairs $(L_j; (-L_j'))$ possesses the same estimate. Consequently, we have proved the following lemma.

Lemma VI.3.1: *The number of conjugate pairs $(L_j; -L_j'))$ originating from the distinct systems $(G_k; \bar{G}_k)$, $k = 1, 2, \ldots, w$, such that the rotations $L_j \to (-L_j')$ are governed by one and the same form (a, b, c) satisfying the conditions*

$$\frac{1}{2} D^{\nu} < \frac{\sqrt{D}}{|a|}, \frac{\sqrt{D}}{|c|} \leqq D^{\nu} \left(0 \leqq \nu \leqq \frac{1}{2}\right) \tag{VI.3.21}$$

possesses the order

$$0(p^{-\frac{s}{2}} D^{\frac{1}{4}} D^{\frac{\nu}{2}} D^{2.3\eta_1}) e_2^{\frac{1}{2}}. \tag{VI.3.22}$$

We turn now to Lemma VI.2.6.

The number of forms $(a, b, c) \in \mathrm{II}$ with given a or c is of the order $0(D^{\xi}\sqrt{e_2})$. Further, we have

$$D^{\frac{1}{2}-\nu} \leqq |a| < 2D^{\frac{1}{2}-\nu}.$$

Consequently, the number of $|a|$ in these conditions is of the order $0(D^{\frac{1}{2}-\nu})$, and the number of $|a|$ for given e_2 has the order $0(D^{\frac{1}{2}-\nu+\xi}e_2^{-\frac{1}{2}})$.

Thus, the total number of forms for such e_2 will be of the order

$$0(D^{\frac{1}{2}-\nu+\xi} \cdot e_2^{-\frac{1}{2}}).$$

According to Lemma VI.3.1, the number of conjugate pairs $(L_j; (-L_j'))$ arising from all such forms will possess the estimate

$$0(p^{\frac{s}{2}} D^{\frac{3}{4}} D^{-\frac{\nu}{2}} D^{2.4\eta_1}).$$

Summing over all $e \mid D$, we discover an estimate for given ν:

$$0(p^{-\frac{s}{2}} D^{\frac{3}{4}} D^{-\frac{\nu}{2}} D^{2.5\eta_1}).$$

The number ν may take on $0(\log D)$ values, so that for all possible cases the estimate

$$0(p^{-\frac{s}{2}} D^{\frac{3}{4}} D^{2.6\eta_1})$$

is valid, and since

$$p^{\frac{s}{2}} = 0(D^{\frac{1}{4}} D^{\frac{1}{2}\eta_2}) = 0(D^{\frac{1}{4}+2.3\eta_1}),$$

we finally obtain the following estimate for the total number of rotations $L_j \to (-L_j')$ for $L_j \in \mathrm{G}_k$, $-L_j' \in \overline{G}_k$, and different k:

$$0(D^{\frac{1}{2}+0.3\eta_1}). \tag{VI.3.23}$$

Set $\eta_3 = 2.5\eta_1; \eta_2 = 4.6\eta_1$. We assumed that the number w of different systems G_k satisfied the condition $w < D^{\frac{1}{2}-\eta_3}$, but this is not possible because of

$$a_1^2 + a_2^2 + \cdots + a_w^2 \geqq \frac{1}{w}(a_1 + a_2 + \cdots + a_w)^2 =$$

$$= \frac{T_2^2}{w} \frac{D^{1-2\eta_1}}{4w} \geqq \frac{1}{4} D^{\frac{1}{2}+\eta_3-2\eta_1} = \frac{1}{4} D^{\frac{1}{2}+0.5\eta_1},$$

which contradicts (VI.3.23).

In this manner we have proved Lemma VI.3.2.

Lemma VI.3.2: *Let η_1 be sufficiently small, $\eta_2 = 4.6\eta_1$, $\eta_3 = 2.5\eta_1$, and let s be chosen such that*

$$D^{\frac{1}{2}+\eta_2} < p^s \leqq p D^{\frac{1}{2}+\eta_2}.$$

We construct the $T_1 \geqq D^{\frac{1}{2}-\eta_1}$ equations

$$l + L_j = X_j \Pi_j \qquad (j = 1, \ldots, T_1),$$

where $L_j \in \mathrm{II}$ and Π_i, L_j, L_j' correspond to (VI.3.1). *Then the number of different Π_j is not less than*

$$D^{\frac{1}{2}-2.5\eta_1} = D^{\frac{1}{2}-\eta_3}.$$

Remark. We note that Lemma VI.3.2 is valid also for the case of rotations $A^{-1}LA$, i.e., for equations having the form

$$l+L_j=\Pi_j X_j .$$

We form the equations

(VI.3.23′) $$l+L_\alpha=X_\alpha\Pi_\alpha(\alpha=1,\ldots,T(D)),\ T(D)=T(\Delta),$$

where

$$\Pi_\alpha=R_{\alpha s}\ldots R_{\alpha i}\ldots R_{\alpha 1},\ \det R_{\alpha i}=p^k,\ L_\alpha\neq L_\beta(\alpha\neq\beta).$$

We choose $T_{\alpha i}=R_{\alpha i}\ldots R_{\alpha 1}$ so that $T_{\alpha i}$ is suitable and $T_{\alpha i}L_\alpha T_{\alpha i}^{-1}\in$ II. If there turns out to be more than one such matrix $T_{\alpha i}$ for given L_α, then we choose $T_{\alpha i}$ in the following manner: let the numbers $\alpha_1,\alpha_2,\ldots,\alpha_n$ be such that $T_{\alpha_1 i},\ldots,T_{\alpha_n i}$ can be chosen in more than one way, and the numbers $\alpha'_1,\ldots,\alpha'_{n'}$ be such that $T_{\alpha'_1 i},\ldots,T_{\alpha'_{n'} i}$ are uniquely determined. Then in all cases where we have a choice for $T_{\alpha_1 i}$, we take one of the $T_{\alpha_1 i}$ which is equal to one of the $T_{\alpha'_j i}$ $(j=1,\ldots,n')$, and if there are no such ones, then $T_{\alpha_1 i}$ is chosen arbitrarily. For the choice of $T_{\alpha_2 i}$ we proceed in the same manner, taking into account that $T_{\alpha_1 i}$ has already been chosen. Let $T_{\alpha_1 i},\ldots,T_{\alpha_m i}$ be already chosen; then we choose $T_{\alpha_{m+1} i}$ from all possible values such that it is equal to one of the $T_{\alpha'_1 i},\ldots,T_{\alpha'_n i},T_{\alpha_1 i},\ldots,\ldots,T_{\alpha_m i}$, and if this is not possible, then $T_{\alpha_{m+1} i}$ is chosen arbitrarily. Thus we produce a choice $T_{\alpha i}$ for all α. In the following we shall deal exclusively with equations of the type (VI.3.23′), where the choice described above has already been performed.

We select from (VI.3.23) T' equations in an arbitrary manner:

(VI.3.24) $$l+L_\beta=X_\beta\Pi_\beta \qquad (\beta=1,\ldots,T'),$$

and we assume that $l^2\equiv D\,(\operatorname{mod} p^{ks})$, $0<l<p^{ks}$, k being a fixed integer; $R^{(1)},\ldots,R^{(g)}$ denotes a complete collection of integral primitive left non-associated matrices with determinant p^k; here $g=(p+1)p^{k-1}$; $\eta_0,\eta_1,\ldots$ is a collection of small positive constants, and s is chosen such that

$$D^{\frac{1}{2}+\eta_2}<p^{ks}<p^kD^{\frac{1}{2}+\eta_2}.$$

Then the following lemma is valid.

Lemma VI.3.3: *Let $r_i(R^{(j)})$ be the number of matrices $T_{\alpha i}$ which are right divisible by $R^{(j)}$. Then, if $T'>T(D)\,D^{-\eta_0}$,*

$$r_i(R^{(j)})\sim T'/g$$

for all $i=1,\ldots,s$ with the possible exception of $\eta(\eta_0)s$ such i and for fixed η_0 and k as $D\to\infty$.

Proof: We give an upper estimate for the number of matrices L for which

$$\frac{\sqrt{D}+b}{|c|}<1+M^{-1}\quad\text{or}\quad\frac{\sqrt{D}+b}{|a|}<1+M^{-1}.$$

Obviously, the right neighbor L_r or the left neighbor L_l to L will satisfy

$$2\frac{\sqrt{D}}{|c_r|} > M \quad \text{or} \quad 2\frac{\sqrt{D}}{|a_l|} > M .$$

Therefore, according to Lemma VI.2.6, the number of such matrices L is of the order

$$0(M^{-1} D^{\frac{1}{2}+\eta}),$$

where η is as small as desired.

We shall say that the matrix L is M-poor, if

$$\max\left(\frac{\sqrt{D}}{|c|}, \frac{\sqrt{D}}{|a|}\right) > \frac{M}{2} \quad \text{or if} \quad \min\left(\frac{\sqrt{D}+b}{|c|}, \frac{\sqrt{D}+b}{|a|}\right) < 1 + M^{-1} .$$

We remove all M-poor matrices L_β from (VI.3.24); their total number is of the order

$$0(D^{\frac{1}{2}+\eta} M^{-1}).$$

Further, we eliminate from (VI.3.24) all matrices L_β for which the matrices

$$L'_\beta = T_{\beta i} L_\beta T_{\beta i}^{-1}$$

are M-poor. The number of such L_β will be of the order

$$0(D^{\frac{1}{2}+\eta} s M^{-1}).$$

We set $M = D^{2\eta_0+\eta} s$; then the number of eliminated matrices is

$$0(D^{\frac{1}{2}-2\eta_0}).$$

For sufficiently large D the number of remaining equations in (VI.3.24) will be larger than $\frac{1}{2} T'$. We assume that in the equations (VI.3.24) the rejection operations described above have been performed, and that the Region II is partitioned into non-overlapping smooth regions whose diameters are less than $D^{\frac{1}{2}} M^{-3}$. Then we find that there exists a region of the partition containing the matrices $L_{\beta_1}, L_{\beta_2}, \ldots, L_{\beta_{T''}}$ such that the inequalities

$$T'' > \frac{T'}{4D} \cdot D M^{-6} = \frac{1}{4} D^{\frac{1}{2}-\eta_0} D^{-12\eta_0 - 6\eta} \geqq D^{\frac{1}{2}-18\eta_0}$$

are satisfied for sufficiently large D. We write down the equations (VI.3.24) for such L_{β_i}:

$$l + L_\beta = X_\beta \Pi_\beta \quad (\beta = 1, \ldots, T''), \quad T'' > D^{\frac{1}{2}-18\eta_0} .$$

We prove that the matrices $T_{\beta i}$ for fixed i are not left associated. Let (we omit some indices)

$$T_1 = \varepsilon \cdot T_2, \; T_1 L_1 T_1^{-1} = L'_1 \in \mathrm{II}, \; \varepsilon \neq \pm 1,$$

and let

$$T_2 L_2 T_2^{-1} = L'_2 \in \mathrm{II}, \; T_2 L_1 T_2^{-1}$$

being an integral matrix. Then according to the choice of T_1 and T_2, $T_2 L_1 T_2^{-1} \in \mathrm{II}$. In other words, the matrices $T_2 L_2 T_2^{-1}, T_1 L_1 T_1^{-1} \in \mathrm{II}$ are M-good; the components of the matrices T_2, T_1 are bounded in absolute value by the number M, and L_1, L_2 are contained in a region whose diameter is less than $D^{\frac{1}{2}} M^{-3}$.

We set

$$T_2 L_1 T_2^{-1} = \begin{pmatrix} b_1' & -a_1' \\ c_1' & -b_1' \end{pmatrix}, \quad T_2 L_2 T_2^{-1} = \begin{pmatrix} b_2' & -a_2' \\ c_2' & -b_2' \end{pmatrix}.$$

Then

$$\| |a_1'| - |a_2'| \|, \ \| |c_1'| - |c_2'| \|, \ \| |b_1'| - |b_2'| \| < D^{\frac{1}{2}} M^{-1},$$

and consequently, if $T_1 L_1 T_1^{-1} \in \mathrm{II}$, then $T_2 L_1 T_2^{-1} \in \mathrm{II}$, which contradicts the above statement. Therefore,

$$l + L_\beta = X_\beta \Pi_\beta \ (\beta = 1, \ldots, T''), \ T'' > D^{\frac{1}{2} - 18\eta_0}$$

and the $T_{\beta i}$ are left non-associated.

We consider now the matrix

$$\text{(VI.3.25)} \qquad \|R_{\beta i}\|, \ \beta = 1, \ldots, T'; \ i = 1, \ldots, s,$$

made up of the matrices $R_{\beta i}$. Suppose that the lemma is false; in this case there are more than $s\xi_0$ indices for which

$$(*) \qquad r_i(R^{(j)}) > \frac{T'}{g}(1 + \xi_0') \quad \text{or} \quad r_i(R^{(j)}) < \frac{T'}{g}(1 - \xi_0').$$

Denote these indices by $i_1, \ldots, i_n$; then we find $s_1 = \dfrac{\xi_0 s}{2}$ indices $i_1, \ldots, i_{s_1}$ among them for which $i_1, \ldots, i_{s_1}$ satisfy one of the inequalities $(*)$ simultaneously. According to the method used in Chapters IV and V, we calculate by columns $i_1, \ldots, i_{s_1}$, and by rows $1, \ldots, T'$ in the matrix (VI.3.25) with the inequalities $(*)$.

In both cases we obtain $T_1 > \eta_1'(\xi_0') T'$ equations of type (VI.3.24) with the following supplementary conditions; in the product Π_α there are more than $(1 + \xi_0'/2) s_1/g$ or less than $(1 - \xi_0'/2) s_1/g$ matrices $R_{\alpha i}$ which are left associated with $R^{(j)}$ for $i = i_1, \ldots, i_{s_1}$. Consequently, according to the discussion at the beginning of the proof, we can find $T' > D^{\frac{1}{2} - 19\eta_0}$ of the equations

$$l + L_\beta = X_\beta \Pi_\beta, \ \beta = 1, \ldots, T'$$

satisfying the conditions:

1) the $T_{\beta i}$ are not left associated;

2) the number of matrices $R_{\beta i}$ left associated with $R^{(j)}$ for $i = i_1, \ldots, i_{s_1}$ in the product $\Pi_\beta = R_{\beta s} \ldots R_{\beta 1}$ will be greater than $(1 + \xi_0'/2) s_1/g$ or less than $(1 - \xi_0'/2) s_1/g$.

By virtue of probabilistic considerations as in Chapters IV and V, we obtain the following upper estimate for the number w' of possible different primitive products $\Pi_\beta = R_{\beta s} \ldots R_{\beta 1}$:

(VI.3.26) $$w' < c(\xi_0, \xi_0')\, w p^{-k s_1 \eta(\xi_0')},$$

(VI.3.27) $$w = (p+1) p^{ks-1}.$$

Applying Lemma VI.3.2 to (VI.3.26), we get a contradiction, which proves the lemma.

Lemma VI.3.4: *For any matrix $R^{(j)}$ taken from the collection $R^{(1)}, \ldots, \ldots, R^{(g)}$, the number $r_1(R^{(j)}, D)$ of equations of the form $l + L_\alpha = X_\alpha R_{\alpha 1}$ ($R_{\alpha 1} = R^{(j)}$) satisfies the asymptotic relation*

$$r_1(R^{(j)}, D) \sim \frac{T(D)}{p^{k-1}(p+1)}$$

as $D \to \infty$.

Lemma VI.3.4 follows immediately from Lemma VI.3.3.

Lemma VI.3.5: *For fixed p, k and as $D \to \infty$,*

$$T_1(p^k, D) \sim 2 \frac{T(D)}{p^{k-1}(p+1)}$$

where $T_1(p^k, D)$ is the number of properly primitive reduced forms with determinant D whose third component is divisible by p^k.

Lemma VI.3.5 follows from Lemma VI.3.4.

Lemma VI.3.6: *For fixed $M \geqq 1$,*

$$T\left(\max\left(\frac{\sqrt{D}}{|a|}, \frac{\sqrt{D}}{|c|}\right) > M\right) < c_1 \frac{T(D)}{\sqrt{M}},$$

where $T\left(\max\left(\frac{\sqrt{D}}{|a|}, \frac{\sqrt{D}}{|c|}\right) > M\right)$ is the number of M-poor matrices and c_1 is an absolute constant.

Proof: We set $M/p \leqq p^{2k} < M$. The number of matrices L for which $\frac{\sqrt{D}}{|a|} > M$ coincides with the number of matrices L for which $\frac{\sqrt{D}}{|c|} > M$. It is sufficient to estimate from above the number of matrices L for which $\frac{\sqrt{D}}{|c|} > M$. We denotes this number by

$$T(M) = T(M, p^k) + T_1(M),$$

where $T(M, p^k)$ is the number of matrices L for which $\frac{\sqrt{D}}{|c|} > M$ and $c \equiv 0 \pmod{p^k}$ and $T_1(M)$ the number of L for which $\frac{\sqrt{D}}{|c|} > M$ and $c \not\equiv 0 \pmod{p^k}$.

It follows from Lemma VI.3.5 that

$$T(M, p^k) < \frac{c_1}{2} \frac{T(D)}{\sqrt{M}}.$$

For an estimate of $T_1(M)$ we consider the transformation

$$ALA^{-1}=L', \ \det A = p^{2k}, \ \frac{\sqrt{D}}{|c|} > M, \ c \not\equiv 0 \,(\operatorname{mod} p^k),$$

and we set

$$A = \begin{pmatrix} p^{n_1} & \xi \\ 0 & p^{n_2} \end{pmatrix}, \ \xi \leqq 0, \ |\xi| < p^{n_2}.$$

Obviously we can find a matrix A such that L' will be integral and properly primitive. We have

$$b' p^{2k} = b p^{2k} - |c\xi|\, p^{n_2},$$
$$c' p^{2k} = c p^{2n_2}.$$

Hence it follows that $c\xi p^{n_2} \equiv 0 \,(\operatorname{mod} p^{2k})$. But $(\xi, p)=1$. In addition, setting

$$c = \bar{c} p^{k_1}, \ k_1 < k,$$

we find

$$k_1 + n_2 \geqq 2k$$

and thus $n_2 > k$. Therefore

$$c' = \bar{c} p^{k_1+n_2-2k} \cdot p^{n_2} \equiv 0 \,(\operatorname{mod} p^k).$$

Moreover we can show that $L' \in \mathrm{II}$. Since

$$\frac{\sqrt{D} \pm b'}{|c'|} = p^{2(k-n_2)} \cdot \frac{\sqrt{D} \pm b}{|c|} \mp \frac{|\xi|}{p^{n_2}},$$

then

$$\frac{\sqrt{D} + b'}{|c'|} > 1, \quad \text{because} \quad \frac{\sqrt{D} + b}{|c|} > M,$$

and

$$\frac{\sqrt{D} - b'}{|c'|} < 1, \quad \text{because} \quad \frac{|\xi|}{p^{n_2}} \leqq 1 - \frac{1}{p^{2(n_2-k)}}, \quad \frac{\sqrt{D} - b}{|c|} < 1.$$

Therefore,

$$T_1(M) < 4 \frac{T(D)}{p^{k-1}(p+1)} < \frac{c_1}{2} \frac{T(D)}{\sqrt{M}},$$

and the lemma above follows.

§ 4. Further Investigation of the Asymptoticity of Hyperbolic Rotations

Lemma VI.4.1: *Suppose $L, L' \in \mathrm{II}$; then*

1) *there exists a matrix* $A = \begin{pmatrix} \alpha & \beta \\ \gamma & \delta \end{pmatrix}$ *with* $\det A = +1$ *and* $\alpha\beta\gamma\delta \leqq 0$ *such that* $ALA^{-1} = L'$;

2) $|\alpha|, |\beta|, |\gamma|, |\delta| \leqq 2 \max\left(\frac{\sqrt{D}}{|a|}, \frac{\sqrt{D}}{|c|}\right) \max\left(\frac{\sqrt{D}}{|a'|}, \frac{\sqrt{D}}{|c'|}\right).$

Proof: Assertion 2) follows immediately from Lemma VI.2.4. We prove 1). We choose $n > 0$ such that

$$(c' - cn)(-a + na') \leqq 0.$$

(obviously, such an n can always be found). Then the matrix

$$A = \begin{pmatrix} \dfrac{b+b'}{\sqrt{d}} & \dfrac{-a+na'}{\sqrt{d}} \\ \dfrac{c'-cn}{\sqrt{d}} & \dfrac{n(b+b')}{\sqrt{d}} \end{pmatrix},$$

where $d=(b+b')^2 n-(-a+na')(c'-cn)$, is sufficient for 1).

We consider the hyperboloid

(VI.4.1) $$x_3^2 - x_1 x_2 = 1,$$

which is said to be normalized and is denoted by H_0. We select the corresponding reduction regions:

I_0: $0<|x_3|<1$, $1-|x_3|<|x_1|<1+|x_3|$,
II_0: $0<x_3<1$, $1-x_3<|x_1|<1+x_3$,
III_0: $0<x_3<1$, $1-x_3<|x_1|<1+x_3$, $x_1<0$,
IV_0: $0<x_3<1$, $1-x_3<|x_1|<1+x_3$, $x_1>0$.

We call the region II_0 fundamental. It is the projection of region II. The points of region II_0 satisfying

$$\sqrt{D}\, x_i \text{ integral } (i=1,2,3),$$
$$(x_1\sqrt{D}, 2x_3\sqrt{D}, x_2\sqrt{D})=1$$

will be called fundamental and admissible.

We consider the fundamental region II_0 and divide it by the lines

(VI.4.2) $$x_3 - x_1 x_2 = 1,\ x_3 = \frac{1}{t} v \quad (v=1,2,\dots,t),$$
$$x_3 - x_1 x_2 = 1,\ x_1 - x_2 = \frac{1}{t} v \quad (v=-[\sqrt{2}\,t],\dots,0,\dots,[\sqrt{2}\,t])$$

(where t is a certain fixed constant) into disjoint regions covering all of II_0.

We select from II_0 the regions with the boundaries

(VI.4.3) $$H_+(k_1): 0<1/x_1 \leqq 2k_1,\ 0<1/-x_2 \leqq 2k_1;$$
$$H_-(k_1): 0<1/-x_1 \leqq 2k_1,\ 0<1/x_2 \leqq 2k_1;$$

and perform the following operation. We eliminate from II_0 all quadrangles in (VI.4.2) which intersect the boundary of $H_+(k_1)$ or $H_-(k_1)$, and we number the remaining quadrangles as follows:

(VI.4.4) $$g_1,\dots,g_{t_1} \in H_+(k_1),\ \bar{g}_1,\dots,\bar{g}_{t_1} \in H_-(k_1).$$

We choose and fix points inside of each of the quadrangles in (VI.4.4) and denote these points by

(VI.4.5) $$0_1,\dots,0_{t_1},\bar{0}_1,\dots,\bar{0}_{t_1}.$$

Let Σ be a region in $H_+(k_1)$ whose boundary is piecewise smooth and does not intersect the boundary of $H_+(k_1)$.

Let $T(g_m)$ $(m=1, \ldots, t_1)$ be the number of fundamental and admissible points in the quadrangle g_m. Applying Lemma VI.3.6 for $M=k_1$, we obtain

$$T(g_1)+\cdots+T(g_{t_1})+T(\bar{g}_1)+\cdots+T(\bar{g}_{t_1}) \geqq \geqq T(D)(1-c_1/\sqrt{k_1}). \tag{VI.4.6}$$

From the numbers $T(g_m)$ we select those for which

$$T(g_m) > \frac{T(D)}{\log^2(D)}. \tag{VI.4.7}$$

Suppose that for a certain m the number $T(g_m)$ satisfies the condition (VI.4.7), and let $L_1, L_2, \ldots, L_{t_m}(t_m = T(g_m))$ be the corresponding vector matrices.

In the following, $\eta_0, \eta_1, \ldots; \varepsilon_0, \varepsilon_1, \ldots$ are small positive constants chosen successively. We fix the number k and thus the power p^k, and select s such that

$$D^{\frac{1}{2}+\eta_2} < p^{ks} \leqq p^k D^{\frac{1}{2}+\eta_2}$$

where $\eta_2 = 4.6\eta_1$, and the value of $\eta_1 > 0$ will be determined later.

We assume that $0 < l < p^{ks}$ and consider the equations

$$l + L_i = X_i \Pi_i, \qquad i = 1, \ldots, t_m. \tag{VI.4.8}$$

Corresponding to these equations we have the unimodular matrices

$$\frac{T_{i1}}{\sqrt{\det T_{i1}}} = A_{i1}, \ldots, \frac{T_{is}}{\sqrt{\det T_{is}}} = A_{is}, \tag{VI.4.9}$$

the number of such matrices being $s \cdot t_m$. In the group of unimodular matrices A we consider the region $\Omega(g_m, \Sigma)$ of unimodular suitable matrices A, which transform the point 0_m of the quadrangle g_m into the region Σ.

We introduce a small number ε_0 (whose exact value will be determined later) and divide the second indices j of the matrices A_{ij} $(j=1, \ldots, s)$ into two types: the index j is of type I if the number of indices i, for which the matrix $A_{ij} \in \Omega(g_m, \Sigma)$ $(i=1, \ldots, t_m)$, lies between the numbers

$$((1-\varepsilon_0)u_m t_m) \quad \text{and} \quad ((1+\varepsilon_0)u_m t_m),$$

where

$$u_m = \frac{\operatorname{meas} \Omega(g_m, \Sigma)}{\operatorname{meas} \Omega(g_m, \Delta)} \in (0, 1);$$

indices j for which this condition is not fulfilled are of type II.

We introduce a small number $\varepsilon_1 > 0$, whose value is also to be fixed later, and we prove that, for sufficiently large D and for the region g_m, the number of type II indices will be less than $\varepsilon_1 s$. We suppose the contrary.

Case 1. There are $s_1 > \varepsilon_1 (s/2)$ indices for which the number of $A_{ij} \in \Omega(g_m, \Sigma)$ $(i=1, \ldots, t_m)$ is greater than $(1+\varepsilon_0)u_m t_m$.

Case 2. There are $s_1 \geqq \varepsilon_1 (s/2)$ indices j for which the number of $A_{ij} \in \Omega(g_m, \Sigma)$ $(i=1, \ldots, t_m)$ is less than $(1-\varepsilon_0)u_m t_m$.

In both cases we have no less than εt_m (ε a small positive constant) first indices i for which the number r_i of operators A_{ij} $(j=j_1, \ldots, j_{s_1})$ with $A_{ij} \in \Omega(g_m, \Sigma)$ does not satisfy the conditions

(VI.4.10) $$(1-\varepsilon_0)u_m t_m \leqq r_i \leqq (1+\varepsilon_0)u_m t_m .$$

We consider $t'_m \geqq \varepsilon t_m$ equations

(VI.4.11) $$l + L_i = X_i R_{is} \ldots R_{i1}, \qquad i=1, \ldots, t'_m$$

with their matrices $T_{ij_1}, \ldots, T_{ij_{s_1}}$ and corresponding unimodular matrices $A_{ij_1}, \ldots, A_{ij_{s_1}}$. We note that if $A_{ij_n} \in \Omega(g_m, \Sigma)$, then the vector matrix $A_{ij_n} L A_{ij_n}^{-1} = L'$ will be fundamental for sufficiently fine partitions of the region III_0 if the distance from the boundary of to the boundary of the region II_0 is not too small. But (see the proof of Lemma VI.3.3), if the matrices $A_{ij_n} \in \Omega(g_m, \Sigma)$, then they are not left associated.

We consider primitive matrices Y_{j_1} with $\det Y_{j_1} = p^{kj_1}$ and their corresponding matrices

$$A_{j_1} = \frac{Y_{j_1}}{\sqrt{\det Y_{j_1}}} \quad (A_{j_1} \text{ suitable}).$$

Let the point $0_m(x_1, x_2, x_3)$ correspond to the vector matrix

$$M = \begin{pmatrix} x_3 & -x_1 \\ x_2 & -x_3 \end{pmatrix}.$$

We investigate the action of the matrix A_{j_1} on M in the sense of the result of the transformation $A_{j_1} M A_{j_1}^{-1}$. First of all we take into account all matrices for which

$$A_{j_1} M A_{j_1}^{-1} \in \Delta \quad (A_{j_1} \text{ suitable}).$$

We fix an arbitrarily large constant k' and consider the region $H_0(k')$; then we obtain the following expression for the number of Y_{j_1} for which $A_{j_1} M A_{j_1}^{-1} \in H_0(k')$ (see Chapter V):

$$f(H_0(k'), p^{kj_1}) = \Psi(p^{kj_1}) \operatorname{meas} \Omega(g_m, H_0(k')) \times (1+\theta\varepsilon_2(k')),$$

where $\varepsilon_2(k') \to 0$ as $k \to \infty$ for given k'. Further, in so far as

$$\frac{\operatorname{meas} \Omega(g_m, H_0(k'))}{\operatorname{meas} \Omega(g_m, \Delta)} = u_m(H_0(k')),$$

we have

$$\operatorname{meas} \Omega(g_m, H_0(k')) = \operatorname{meas} \Omega(g_m, \Delta)\,(1+\theta\varepsilon_3(k')),$$

where $\varepsilon_3(k') \to 0$ as $k' \to \infty$. Hence, taking into account that

$$\Psi(\Delta, p^{kj_1}) = \frac{6}{\pi^2}\, p^{kj_1 - 1}(p+1)\left(1 + \frac{\theta}{\log^{40} p^{kj_1}}\right),$$

we find that

(VI.4.12)
$$f(\Delta, p^{kj_1}) = \frac{6}{\pi^2} \operatorname{meas} \Omega(g_m, \Delta) p^{kj_1-1}(p+1) \times \times (1+\theta\varepsilon_4(k,k')).$$

In other words, the expression

$$f(\Delta, p^{kj_1}) \sim 2p^{kj_1-1}(p+1)$$

is valid for $f(\Delta, p^{kj_1})$. Comparing this with (VI.4.10), we find that

(VI.4.13)
$$\frac{6}{\pi^2} \operatorname{meas} \Omega(g_m, \Delta) = 2.$$

Now we consider left non-associated primitive matrices Y_{j_1} for which $A_{j_1} \in \Omega(g_m, \Sigma)$; according to the conclusions just drawn we see that the number of such matrices is

$$f(\Omega(g_m, \Sigma), p^{kj_1}) \sim \frac{6}{\pi^2} \operatorname{meas} \Omega(g_m, \Sigma) p^{kj_1-1}(p+1),$$

or

$$\tfrac{1}{2} f(\Omega(g_m, \Sigma), p^{kj_1}) \sim u_m(\Sigma)\, w(p^{kj_1}),$$

where

$$w(p^{kj_1}) = p^{kj_1-1}(p+1).$$

Let the matrix T_{ij_1} be fixed in the equations (VI.4.11). We now pass to the matrix T_{ij_2}. We shall find an expression for the number of matrices T_{ij_2} which may occur in the equations (VI.4.11) and for which $A_{ij_2} \in \Omega(g_m, \Sigma)$, under the condition that the matrix $S_{ij_2} = T_{ij_2} T_{ij_1}^{-1}$ is primitive and not left divisible by R_{ij_1} with determinant p. We have

$$S_{ij_2} \cdot p^{-\frac{j_2-j_1}{2}} \in \Omega'(g_m, \Sigma) = \Omega', \quad \operatorname{meas} \Omega = \operatorname{meas} \Omega'.$$

In view of this, the required number will be

$$\tfrac{1}{2}\left[f(\Omega', p^{(j_2-j_1)k}) - f(\Omega', p^{(j_2-j_1)k}, R_{ij_1})\right] = u_m p^{(j_2-j_1)k}(1+\theta\varepsilon_4(k,k')).$$

The number of those T_{ij_2} which may occur in (VI.4.11) and for which $A_{ij_2} \in \Omega(g_m, \Sigma)$ is equal to

$$(1-u_m) p^{(j_2-j_1)k}(1+\theta\varepsilon_4(k,k')).$$

We pass over to the matrix T_{ij_3} for fixed T_{ij_2} and perform exactly the same calculations for the indices $j_3, \ldots, j_{s_1}$.

In Chapter V an upper estimate was given for the number of distinct $T_{ij_{s_1}}$, k' and k in $\varepsilon_4(k,k')$ being chosen so that

$$2\varepsilon_4(k,k') < \varepsilon_5 \cdot \tfrac{1}{4},$$

where ε_5 depends only on ε_0.

Further, since

$$(1+\theta_1\varepsilon_4) \ldots (1+\theta_{s_1}\varepsilon_4) < \exp(2s_1\varepsilon_4(k,k')),$$

the number w_1 of possible different values for $T_{ij_{s_1}}$ $(i=1, \ldots, t'_m)$ in the equations (VI.4.11) receives the estimate

$$w_1 \leqq w(p^{kj_{s_1}}) \exp(-\tfrac{3}{4}\varepsilon_5 s_1).$$

For each given value of $T_{ij_{s_1}}$, the integral matrix T_{is} may take on no more than $p^{k(s-j_{s_1})}$ different left non-associated values, so that the total number of different T_{is} in the equations (VI.4.11) can not exceed

$$p^{ks}(p+1)\exp(-\tfrac{3}{4}\varepsilon_5 s_1) = (p+1)p^{ks-\frac{3}{4\log p}s_1\varepsilon_5}.$$

Since $s_1 \geqq \varepsilon_1 \frac{s}{2}$, then, setting $\varepsilon_6 = \varepsilon_1 \varepsilon_5 \frac{1}{4k\log p}$, we find that, for sufficiently large s, the following upper estimate for the number w_2 of possible distinct T_{is} is valid:

$$w_2 < p^{ks(1-\varepsilon_6)}.$$

Turning now to the Lemma VI.3.2 for $t'_m > \frac{\varepsilon(\varepsilon_0)}{\log^2 D} T(D)$, we arrive at a contradiction.

Remark. It is easy to see that all of our lemmas for rotations of the form ALA^{-1} are also true for rotations of the form $A^{-1}LA$. Therefore for the equation

$$l + L_\alpha = \Pi_\alpha X_\alpha$$

the discussion in § 2 and § 3 is valid. We make use of this in the following.

Let a region $\Sigma'' \subsetneqq \mathrm{IV}_0$ be given. We consider the transformation $AM'A^{-1}$, where A is suitable and $\det A = +1$, and where the vector matrix

$$M' = \begin{pmatrix} x'_3 & -x'_1 \\ x'_2 & -x'_3 \end{pmatrix}$$

corresponds to the point (x'_1, x'_2, x'_3) of the region IV_0, where the coordinate x'_3 is less than every x''_3 (x''_3 being a coordinate of a point on the region Σ''). Then in the quaternary region $\Omega(M', \Sigma'')$, where the A are all suitable, there are no two left associates. This follows immediately from § 2.

We consider regions $\Sigma', \Sigma'' \subsetneqq \mathrm{IV}_0$ under the conditions

1) $V(\Sigma') = V(\Sigma'')$;

2) x'_3 (a coordinate of a point in Σ') is less than every x''_3 (coordinates of points on Σ'');

3) $d'd'' < \xi$, where d', d'' are the diameters of thr regions Σ' and Σ'' respectively and ξ is a small positive quantity;

4) the regions Σ', Σ'' are connected and the lengths of their boundaries are of the order $0(\xi)$.

We denote by $\bar{\Omega}(M, \Sigma)$ the quaternary region which contains all suitable matrices A for which $AMA^{-1} \in \Sigma$, and by $\Omega(M, \Sigma)$ our arbitrary connected region which contains only left non-associated suitable matrices A such that $AMA^{-1} \in \Sigma$.

For instance, in the case of the regions Σ', Σ'', we have

$$\Omega(M', \Sigma'') = \bar{\Omega}(M', \Sigma'').$$

This implies that for any $M_1', M_2' \in \Sigma', M_1'', M_2'' \in \Sigma''$ we obtain

(VI.4.14)
$$\begin{gathered}
\frac{\operatorname{meas}\Omega(M_1',\Sigma'')}{\operatorname{meas}\Omega(M_2',\Sigma'')} = 1+\theta'\eta(\xi), \quad |\theta'|<1,\\
\frac{\operatorname{meas}\Omega(M_1'',\Sigma')}{\operatorname{meas}\Omega(M_2'',\Sigma')} = 1+\theta''\eta(\xi), \quad |\theta''|<1,\\
\frac{\operatorname{meas}\Omega(M_2'',\Sigma')}{\operatorname{meas}\Omega(M_2',\Sigma'')} = 1+\bar{\theta}''\eta(\xi), \quad |\overline{\theta''}|<1,\\
\frac{\operatorname{meas}\Omega(M_1'',\Sigma')}{\operatorname{meas}\Omega(M_1',\Sigma'')} = 1+\theta\eta(\xi), \quad |\theta|<1,\\
(\eta(\xi)\to 0 \quad \text{as} \quad \xi\to 0).
\end{gathered}$$

From Lemma V.11.1 it follows immediately that for any j_1

(VI.4.15)
$$\sigma = \frac{f(\Omega(\Sigma',\Sigma''),p^{kj_1})}{f(\Omega(\Sigma'',\Sigma'),p^{kj_1})} = \frac{\operatorname{meas}\Omega(\Sigma',\Sigma'')}{\operatorname{meas}\Omega(\Sigma'',\Sigma')}\times(1+\eta(p^k,D))$$
$$(\eta(p^k,D)\to 0 \quad \text{as} \quad D\to\infty).$$

Fixing ξ, we set $\varepsilon_1 = 1/t_1^2$ and choose t_1 sufficiently large. Consider all $T'(g_m)\in\Sigma'$ and $T''(g_m)\in\Sigma''$. We write the corresponding equations

(VI.4.16) $$l+L_\alpha' = \Pi_\alpha' X_\alpha', \quad L_\alpha'\in\Sigma',$$

(VI.4.17) $$l+L_\alpha'' = X_\alpha''\Pi_\alpha'', \quad L_\alpha''\in\Sigma''.$$

Further, we choose j_1 and denote by r_i the number of second indices α in the equations (VI.4.16) for which the matrix $T_{\alpha j_1}$ $(\alpha=1,\dots,r_i)$ fulfills $T_{\alpha j_1}^{-1}L_\alpha' T_{\alpha j_1}\in\Sigma''$, and we denote by $\bar{r}_r$ the number of second indices α in the equations (VI.4.17) for which the matrix $T_{\alpha j_1}$ $(\alpha=1,\dots,\bar{r}_r)$ satisfies $T_{\alpha j_1}L_\alpha'' T_{\alpha j_1}^{-1}\in\Sigma'$, where in this case left associates are taken into account.

Then we get $r_i=\bar{r}_r$,

$$r_i = T(\Sigma')\frac{\operatorname{meas}\Omega(\Sigma',\Sigma'')}{\operatorname{meas}\Omega(\Sigma',\Delta)}(1+\eta(\xi,p,D)),$$
$$\begin{aligned}
\bar{r}_r &= \sigma\cdot T(\Sigma'')\frac{\operatorname{meas}\Omega(\Sigma'',\Sigma')}{\operatorname{meas}\Omega(\Sigma'',\Delta)}(1+\eta_1(\xi,p,D))\\
&= T(\Sigma'')\frac{\operatorname{meas}\Omega(\Sigma',\Sigma'')}{\operatorname{meas}\Omega(\Sigma'',\Delta)}(1+\eta_2(\xi,p,D)),
\end{aligned}$$

and since $\dfrac{\operatorname{meas}\Omega(\Sigma',\Delta)}{\operatorname{meas}\Omega(\Sigma'',\Delta)}\sim 1$, then

$$T(\Sigma') = T(\Sigma'')(1+\eta_3(\xi,p,D)),$$
$$\eta_3(\xi,p,D)\to 0 \quad \text{as} \quad D\to\infty,\ \xi\to 0$$

and for fixed p.

Theorem VI.2.1 is proved.

§ 5. An Ergodic Theorem and a Mixing Theorem

In Chapter IV an ergodic theorem was proved for the spherical case. In the case of the hyperboloid of one sheet difficulties arise in the choice of an abstract flow. From the preceding it is clear that the rotation $RLR^{-1}=L'$, L, $L'\in \mathrm{II}$ is realized with a set of matrices of the form $\pm R\cdot\varepsilon^{-n}$, where $\varepsilon=T+UL$, $\det\varepsilon=1$, and n runs through all integers from $-\infty$ to $+\infty$. In addition to this, even if we impose on the matrix R the condition of suitability, this does not produce uniqueness.

Therefore the same choice (generally speaking) of suitable R for the rotation RLR^{-1} is not unique.

We stress that the proof of Theorem VI.2.1 was possible because R was contained in a suitable region. We denote this region by ω; then it is obvious that Theorem VI.2.1 can be proved in the same manner if $R\in\omega\cdot\varepsilon$, where ε is a 2×2-matrix with $\det\varepsilon=1$.

Moreover, generally speaking, we would not be able to choose any integral domain ω' of matrix operators such that the above non-uniqueness would be removed. This turns out to be a consequence of the fact that the number of representations of a class in cycles is not identical for different classes. Hence it follows that the choice of an abstract flow is "performed", but not given canonically as in the case of the sphere.

Let

$$l+L=XR_sR_{s-1}\dots R_1$$
$$T_i=R_i\dots R_1$$
$$\det R_i=p^k.$$

Then $S_j=T_{j+1}T_j^{-1}$ $(j=0,1,\dots,s-1)$ may be used as matrix operators to induce an abstract flow; T_i is chosen as in § 3. We denote the operation of type RLR^{-1} by A and the image L' of L by AL; the n-fold operation is designated by $A^{(n)}$ and the corresponding images by $A^{(n)}L$.

Further, by $f_\Sigma(A^{(n)}L)$ we denote the characteristic function of a set of primitive integral points; this function is defined by the equation

$$f_\Sigma(A^{(n-1)}L)=\begin{cases}0 & \text{if } A^{(n)}L\notin\Sigma\\ \dfrac{V(\Sigma)}{V(\Delta)}\quad\dfrac{1}{\operatorname{meas}\Omega(L,\Sigma)} & \text{if } A^{(n)}L\in\Sigma.\end{cases}$$

Using § 4 it is then easy to prove the following theorem.

Theorem VI.4.1 *(ergodic theorem):*

$$\frac{1}{s}\sum_{n=1}^{s} f_\Sigma(A^{(n-1)}L)=\frac{V(\Sigma)}{V(\Delta)}(1+\eta(p^k,\Sigma,D))$$

for all primitive points L with the possible exception of $o(T(\Delta))$ such points; here $s\geqq c_0\log D$ and $\eta(p^k,\Sigma,D)\to 0$ for fixed p^k, Σ as $D\to\infty$.

Suppose N is an arbitrary set of primitive integral points contained in the region II. We denote by $A^{(n)}N$ the set of primitive integral points of region II obtained by the n-fold transformation of N by A; by $M(N)$ we mean the number of points in N; by $M(A^{(n)}N \cap \Sigma)$ the number of points of $A^{(n)}N$ lying in the region Σ, where if m of the points can be translated to the region Σ from one point with suitable matrices R, then the number of these is assumed to be equal to m. Then a more general distribution theorem is valid, which turns out to be a generalization of Theorem VI.2.1.

Theorem VI.4.2 *(mixing theorem): Let* $l = 0, 1, 2, \ldots, s > c_1 \log D$ *and* $M(N) > \varepsilon_0 T(\Delta)$ *(*ε_0 *being a fixed number). Then for all indices* l *with the possible exception of* $s \cdot o(1)$ *such indices, we have*

$$M(A^{(l)}N \cap \Sigma) = \frac{V(\Sigma)}{V(\Delta)} M(N)\,(1 + \eta(p^k, \Sigma, \varepsilon_0, D)),$$

where $\eta(p^k, \Sigma, \varepsilon_0, D) \to 0$ *for given* $p^k, \Sigma, \varepsilon_0$ *as* $D \to \infty$.

We remark that the proof of Theorem VI.2.1 depends upon an estimate of the value of $T(\Delta)$. It is known that

$$\log T(\Delta) \sim \tfrac{1}{2} \log D\,.$$

We have made use of this in proving Theorem VI.2.1. Theorem VI.4.2 is also easy to prove in this way.

Chapter VII

Algebraic Fields of a More General Type

§ 1. General Remarks

In Chapter VI we encountered the basic difficulties occurring in the study of ergodic properties of fields of a more general type than imaginary quadratic. The appearance of infinitely many units, the problems concerning regularity behavior of the field, which connects the number of ideal classes with the number of in a given sense "reduced" representation of the ideal classes, the totality of these reduced representations – each of these leaves its imprint on the corresponding asymptotic and ergodic properties. At the same time, aside from the ergodic properties, we are also interested in asymptotic theorems in conducting the analysis; for instance, the theorem on cycles by B. F. SKUBENKO (Theorem VI.2.2).

We turn now to a general formulation of the problem. Let a sequence of irreducible equations of the type

(VII.1.1) $$L^n + a_1(D)L^{n-1} + \cdots + a_n(D) = 0$$

be given, where $a_1(D), \ldots, a_n(D)$ are integral functions of the parameter D, for which as $D \to \infty$,

(VII.1.2) $$a_k(D) = 0(D^{\frac{k}{n}}).$$

They determine a sequence of algebraic fields. An example for such a system is the sequence of equations

(VII.1.3) $$L^n = \pm D; \quad D \to \infty$$

for $n = 2, 3, \ldots$, D not divisible by an nth power, determining the fields $k(\sqrt[n]{D})$.

We shall consider solutions of equations of the type VII.1.1 in $n \times n$-matrices. Such a formulation is easily connected with the question of representation of fields by integral matrices, which will be discussed later.

We consider the matrix

(VII.1.4) $$L = \left\| \begin{matrix} x_{11} & \cdots & x_{1n} \\ \vdots & \cdots & \vdots \\ x_{n1} & \cdots & x_{nn} \end{matrix} \right\|$$

with real coefficients and satisfying the equation (VII.1.1).

It contains n^2 integral parameters. The equation (VII.1.1) will generally speaking impose n conditions on them, and the matrix L will depend on $n^2 - n$ independent parameters. We remark that together with the matrices L, the matrices $L'=SLS^{-1}$ for any non-singular matrix S will satisfy our equation (VII.1.1). In the simplest case of a quadratic field ($n=2$ in the equations (VII.1.3)), the general form of the matrix L, depending on $2^2-2=2$ parameters, can be written down immediately. For those equations the general form of the matrix solution is

$$L=\begin{Vmatrix} b & -a \\ c & -b \end{Vmatrix}, \quad \text{where} \quad b^2-ac=\pm D,$$

so that we have an algebraic dependence on two parameters. For an integral matrix L the minimal polynomial of L coincides with the left side of (VII.1.1), which is an irreducible polynomial. In view of this, the coefficients $a_1(D), a_2(D), \ldots, a_n(D)$ must be basic symmetric functions of the elements of the matrix L satisfying the equation (VII.1.1). In particular, $\mathrm{Sp}(L)=a_1(D)$; $\det L=(-1)^n a_n(D)$. Therefore the integral matrix (VII.1.4) generates a system of n diophantine equations of the type:

$$\text{(VII.1.5)} \qquad \begin{cases} \varphi_1(x_{ij})=a_1(D) \\ \varphi_2(x_{ij})=a_2(D) \\ \quad \vdots \qquad\quad \vdots \\ \varphi_n(x_{ij})=a_n(D), \end{cases}$$

where $\varphi_k(x_{ij})$ $(k=1,2,\ldots,n)$ are homogeneous polynomials of degree k in the n^2 variables x_{ij}, differing only in sign from the respective coefficient of the "secular" equation of the matrix L, so that, for example,

$$\varphi_1(x_{ij})=-(x_{11}+x_{22}+\cdots+x_{nn}).$$

The object of our investigations will be the asymptotic geometry of solutions to the system of equations (VII.1.5), that is, the problem of representation of integral points in different regions on intersections with algebraic surfaces (the algebraic manifold defined by equations (VII.1.5)). In addition to this, we investigate questions as to whether there are solutions on the given regions satisfying prescribed congruences modulo a given q; for example, whether there exist solutions such that x_{11} and x_{12} are divisible by q, etc.

Of course, the diophantine equations (VII.1.5) have a special form. However, their properties are obviously of interest for the following calculations.

First of all, the study of asymptotic solutions of a general system of n diophantine equations with n^2 variables is inaccessible by the usual contemporary methods of analytic number theory, so that it is expedient to investigate special systems which are conformable with these methods. Even the special case $n=2$, examined in Chapters IV, V, and VI (with generalizations given in the monograph of A. V. Malyshev), reduces to the task of determining asymptotic representations of numbers by determinate and indeterminate ternary quadratic forms, which is inaccessable by the usual methods of analytic number theory.

Second, the solutions of the system of equations (VII.1.5) are undoubtedly connected with the behavior of the ideal classes of the algebraic field, given by the equation (VII.1.1). This connection was particularly clear in Chapters IV and V, where the number of solutions on the corresponding surfaces turned out to be asymptotically proportional to the number of ideal classes of the corresponding imaginary quadratic field (see, for example, the relation (V.2.2)). Therefore we must obtain information on the ideal classes of the respective algebraic fields in order to investigate these equations.

Third, in the investigation of solutions of the equations (VII.1.5), phenomena of ergodic character are naturally uncovered, producing certain properties of sequences of algebraic fields with the same degree. These properties of ergodic character are of independent interest.

In the following points we touch upon some of these problems; however, in so far as work in this region has hardly begun, we restrict ourselves to basic short explanations without proofs.

§ 2. On the Representations of Algebraic Numbers by Integral Matrices*

We consider an algebraic field k_D, generated from the field of rational numbers by the equation (VII.1.1). We shall investigate isomorphic representations of integral numbers $\alpha \in k_D$ by integral $n \times n$-matrices. Such representations we denote by $D(\alpha)$. Two representations $D_1(\alpha)$ and $D_2(\alpha)$ will be assumed equivalent, if

(VII.2.1) $$D_2(\alpha) = \varepsilon D_1(\alpha) \varepsilon^{-1},$$

where ε is an integral unimodular matrix ($\det \varepsilon = \pm 1$). Equivalent representations compose a class of representations. The number of classes is closely connected with the number of ideal classes of k_D. The following theorem is valid, and was proved by A. Chatelet and in a more general form by J. Schur. We state it in the following manner.

Theorem VII.2.1: *The number of classes of representations $\{D(\alpha)\}$ coincides with the number of ideal classes of k_D.*

Each fixed representation can be obtained from the "multiplication table" of the integers of k_D by the fixed basis number $\{\bar{a}\}$ of a certain ideal a, since, in view of Theorem VII.2.1, they correspond to the ideal classes (see, for instance, Chebotarev [1]). The transition from one basis $\{\bar{a}\}$ to another corresponds to the transition between equivalent representations (VII.2.1). In view of this a given representation $D(\alpha)$ can be written in the form $A_{\{\bar{a}\}}(\alpha)$. From each representation $A_{\{\bar{a}\}}(\alpha)$ we can

* Remarks by D. K. Faddeev have been useful to me in handling the representations of algebraic numbers by matrices.

pass over to another representation $A_{\{\bar{b}\}}(\alpha)$ by means of the "rotation"

(VII.2.2) $$A_{\{\bar{b}\}}(\alpha) = P A_{\{\bar{a}\}}(\alpha) P^{-1},$$

where P is a prime matrix with $\det(P) = p = \mathrm{Norm}(\pi)$, π being a prime ideal of first degree contained in the class of ideals represented by ba^{-1}, so that $\pi \sim ba^{-1}$ ($\sim$ means equivalence of ideals). Conversely, if P is a prime matrix which transforms according to the formula (VII.2.2) the representation $A_{\{\bar{a}\}}(\alpha)$ into the representation $A_{\{\bar{b}\}}(\alpha)$, then $b \sim a\pi$, where π is a prime ideal connected with P in the manner described above.

In this sense we can consider a "representation flow" carrying representations corresponding to the ideal a with basis $\{\bar{a}\}$ into ideal classes with representatives $a\pi, a\pi^2, a\pi^3, \ldots$, using a prime matrix P as an operator. This is what was done in Chapters IV and V for the case of quadratic fields.

It is possible to explain and supplement the considerations above in the following manner (see LINNIK [5]).

If $K(\eta)$ is a field of nth degree and $[\mu_0, \mu_1, \ldots, \mu_{n-1}]$ is a basis of an ideal of the field, θ an integral number of the field, then for the one row matrices $\|\mu_j\|$ and $\|\theta\mu_j\|$ we have the relation

$$\|\theta\mu_j\| = \Omega\,\|\mu_j\|,$$

where $\Omega = \|a_{ij}\|$ is an integral $n \times n$-matrix. Let $[\omega_0, \omega_1, \ldots, \omega_{n-1}]$ be a fundamental basis of the field; for each ideal in the field we form for the numbers $\omega_0, \omega_1, \ldots, \omega_{n-1}$ the respective matrices $\Omega_0, \Omega_1, \ldots, \Omega_{n-1}$ (see CHEBOTAREV [1]). Two such systems of matrices will be said to be unimodularly similar if one of them goes into the other by right multiplication of ε and left multiplication of ε^{-1}, where ε is an integral matrix unit ($\det(\varepsilon) = \pm 1$). In this manner, the systems of matrices $\Omega_0, \Omega_1, \ldots, \Omega_{n-1}$ are divided into classes.

The theorem of CHATELET-SCHUR says then that the number of such classes is equal to the number of ideal classes of the field.

We can supplement this theorem. The transition from one system of matrices $[\Omega_0, \Omega_1, \ldots, \Omega_{n-1}]$ to another system $[\Omega'_0, \Omega'_1, \ldots, \Omega'_{n-1}]$ belonging to another class can be accomplished, generally speaking, with the help of a similarity transformation by an integral non-unimodular matrix Q such that $\Omega'_i = Q^{-1}\Omega_i Q$. Here the ideal class characterized by the system $[\Omega'_0, \Omega'_1, \ldots, \Omega'_{n-1}]$ is obtained from the ideal class characterized by the system $\Omega_0, \Omega_1, \ldots, \Omega_{n-1}$ by multiplying it with one of the ideal classes obtained by factoring the determinant of the matrix Q into prime ideal factors in the field $K(\eta)$. The above property is closely connected with the following property of integral matrices.

Theorem VII.2.2: *If L and Q are integral matrices such that* $\det(Q)$ *is free of squares, then, in order for the matrix $L' = Q^{-1}LQ$ to be integral, it is necessary and sufficient that there exist an integer l such that*

(VII.2.3) $$lE + L = QU,$$

where $E = \|\delta_{ij}\|$ *(δ_{ij} is the Kronecker delta symbol) and* U *is an integral matrix.*

The "traveling matrix" Q may be compared to one of the "traveling ideals": $a = (\det(Q), l + \eta)$, where η is one of the roots of the minimal equation which the matrix satisfies.

In the following we shall write 1 instead of the symbol E; the matrix lE is written as the number l. In this more general case we obtain thus the equation

$$l + L = QU$$

employed in the preceding chapters, and we may speak of rotations (L, L') with the respective equations:

$$l + L = QU\,; \quad l + L' = UQ\,; \quad Q^{-1}LQ = L'\,; \quad \det(Q) = q\,. \tag{VII.2.4}$$

It is true that $\det(Q)$ will be free of squares, but earlier we employed a matrix Q with $\det(Q)$ being a power of a prime, and here the result is somewhat more complicated. However, this difficulty can be overcome, and we do not go into it here.

We remark yet that for the consideration of flows of the above type we shall use ideal classes. However, in contrast to the case of an imaginary quadratic field presented in Chapters IV and V, the result here will be in keeping with those in Chapter VI, that is the "reduction" in a certain sense of representatives of the ideal classes (analogous to the reduction of forms in Chapter VI) will be equivalent. Up until now, for instance, it is not known whether there exists an infinite set of real quadratic fields with class number $h = 1$; this problem, as is known, is connected with the problem of possible growth of the field regulator. In the case $h = 1$ all representations of ideal classes are equivalent. Nevertheless, the investigation of asymptotic solutions of the equations (VII.1.5) is of interest, since it includes, for example, a problem concerning representation of ideals with small norms. Let us consider, for example, the Kummer field corresponding to equation (VII.1.3); let $d = d(D)$ be its discriminant. It is known from the classical theory of algebraic numbers that in each ideal class there is an ideal a with $\mathrm{Norm}(a) \leqq \sqrt{|d|}$.

Similar to Chapter V, where we deduced the asymptotic distribution of ideals having norms not exceeding $\varepsilon \sqrt{D}$, ε being an arbitrarily small fixed number, here we can study the asymptotic distribution of ideals with norms not exceeding $\varepsilon \sqrt{D}$, and, putting together matrix representations with fixed ideal bases as was described above, we can investigate the asymptotic geometry of such representations as in Chapters IV, V, and VI. The measure here turns out to be the Haar measure on the group of unimodular matrices.

We consider the problem of translation to possibly more general cases of constructions, in order to obtain the asymptotic formulae of

Chapters IV, V, VI. Here we refer to the theorems on rotations, the analogy of versor rays, and lemmas of the type of Lemma V.11.1 concerning the distribution of integral matrices with large determinants.

§ 3. Rotations

We give first some asymptotic properties of "rotations" (L, L') of two solutions L and L' to the equation (VII.1.1). Taking into account condition (VII.1.2), we say that the solution L is normally proportioned if $L=0(D^{\frac{1}{n}})$ as $D\to\infty$. We remark that then $|\det L|=0(D)$. By this notation we mean that all coefficients of L have the order $0(D^{\frac{1}{n}})$. In general we say that a matrix M with $|\det M|$ increasing beyond all bounds is normally proportioned, if

$$M=0(|\det M|^{\frac{1}{n}}).$$

By virtue of condition (VII.1.2) two determinations of L as normally proportioned matrices agree among themselves.

Considering a sequence of equations of type (VII.1.1), we select from them those for which $\theta_1, \theta_2, \ldots, \theta_n$ are different and "well distributed" in the following sense:

(VII.3.1)
$$c_1 D^{\frac{1}{n}} \geqq |\theta_a-\theta_b| \geqq c_2 D^{\frac{1}{n}}.$$

Such sequences of equations are called well separated. Other sequences will not be considered. The equations (VII.1.3) serve as a typical example for a well distributed sequence of equations generating the Kummer field; in particular equations of the type $L^2=D(D\gtrless 0)$ considered in Chapters III, IV, V.

Let L and L' be two solutions of the well distributed equations (VII.1.1) for sufficiently large D. We assume that L' is relatively close to L. By this we understand that the differences between the coefficients of L and L' do not exceed

$$\zeta D^{\frac{1}{n}}$$

where ζ is a sufficiently small number. We prove that in this case there exists a matrix S such that: $|\det S| \asymp (\det L)^{n-1}$, S is normally proportioned:

$$S=0(|\det S|^{\frac{1}{n}}),$$

and

$$SLS^{-1}=L'.$$

Here the symbol "$\asymp$" is the Hardy equivalence symbol: $A(D)\asymp B(D)$ as $D\to\infty$ if the ratio $\frac{A(D)}{B(D)}$ is bounded from above and below by positive constants.

For this we use a formula of J. J. Sylvester. Both of our matrices L and L' fulfill equation (VII.1.1), which we write as $f(L)=0$. We now construct the polynomial in two variables

(VII.3.2)
$$\Psi(x, x')=\frac{f(x)-f(x')}{x-x'},$$

writing it in the form

(VII.3.3) $$\Psi(x, x') = \sum_{k=1}^{n-1} \sum_{i+j=k} a_{ij} x^i x'^j ,$$

and we consider the matrix $\Psi = \Psi(L, L')$ obtained by substituting the matrices L and L' for x and x' respectively in the right side of (VII.3.3). An immediate calculation shows that

(VII.3.4) $$\Psi L = L' \Psi .$$

We consider further the behavior of det Ψ. In view of the properties of L and L' we may write

(VII.3.5) $$L' = L + \tau D^{\frac{1}{n}}$$

where τ is a sufficiently small matrix.

We deduce without difficulty from (VII.3.2)

(VII.3.6) $$\Psi = \Psi(L, L') = f'(L) + \tau_1 D^{\frac{n-1}{n}} ,$$

where τ_1 is a small matrix. The property of distribution (VII.3.1) shows us that

$$|\det f'(L)| \asymp D^{n-1} ,$$

so that from (VII.3.6) we conclude

(VII.3.7) $$|\det \Psi(L, L')| \asymp D^{n-1} .$$

Further, from (VII.3.2) and from the fact that L and L' are normally proportioned we deduce

(VII.3.8) $$\Psi = \Psi(L, L') = 0(|\det \Psi|^{\frac{1}{n}}) ,$$

so that the matrix Ψ is normally proportioned.

Therefore, with the help of matrices of the type Ψ, normally proportioned, we can perform rotations in the neighborhood of the given solution L of equation (VII.1.1) in the manner described above. Of course we also find that in general it is possible to go over from one normally proportioned solution L to another with the help of a normally proportioned matrix. It turns out to be possible to construct for the matrix Ψ a normally proportioned matrix giving the same rotation and having a significantly smaller determinant. In fact, we can construct a matrix U such that

$$U L U^{-1} = L', \ \det U D^{\frac{n-1}{2}} \quad , \ U = 0(D^{\frac{n-1}{2n}}) .$$

This corresponds to those matrices giving effect to the rotations which we considered in Chapters IV, V and VI; there they arose from the theory of reduction of binary quadratic forms. This assertion is also connected with the classical theorem that, in each class of ideals of the field k with discriminant d, it is possible to select an ideal with norm not exceeding $\sqrt{d}$.

We consider now the construction corresponding to the versor rotation studied in the preceding chapters. Versor rotations (L, L') were introduced there for 2 × 2-matrices satisfying the equation $L^2 = D (D \gtrless 0)$, for which the relations

$$\text{(VII.3.9)} \qquad I + L = \Pi V; \quad I + L' = \bar{\Pi} V'; \quad \det \Pi \asymp D^{\frac{1}{2} + \tau}$$

were valid. Here, $\tau > 0$ is a small constant and $\bar{\Pi}$ is the conjugate matrix to Π. The matrices C giving rise to the versor rotations $CLC^{-1} = L'$ enjoyed the convenient property

$$\text{(VII.3.10)} \qquad 2\,\mathrm{Sp}(C\bar{\Pi}) \equiv 0 (\bmod \det \Pi)\,.$$

Here it was important that the discussion concerned 2 × 2-matrices for which the conjugate matrix $\bar{\Pi}$ was simply related to the initial matrix Π; in particular, if Π was normally proportioned, so that $\Pi = 0(|\det \Pi|^{\frac{1}{2}})$, then $\bar{\Pi} = 0(|\det \Pi|^{\frac{1}{2}})$, since $\det \Pi = \det \bar{\Pi}$. For matrices of higher order this is not so.

For an expedient definition of the concept of versor rotation for matrices L of higher order, we remark that in the case of 2 × 2-matrices we could have used instead of the matrix Π the matrix Π^T, that is, the transpose of Π. In fact, let $\Pi = \begin{pmatrix} \alpha & \beta \\ \gamma & \delta \end{pmatrix}$; then $\bar{\Pi} = \begin{pmatrix} \delta & -\beta \\ -\gamma & \alpha \end{pmatrix}$. Setting

$$i = \begin{pmatrix} 0 & -1 \\ 1 & 0 \end{pmatrix},$$

it is easy to show that

$$\bar{\Pi} = i \Pi^T i^{-1},$$

so that operating with the matrix Π^T instead of the matrix $\bar{\Pi}$ would not have caused difficulty.

At the same time, $\det \Pi^T = \det \Pi$ for matrices of any order. Therefore it is natural to define a versor rotation as a rotation corresponding to the relations

$$\text{(VII.3.11)} \qquad I + L = \Pi V; \quad I + L' = \Pi^T V'\,.$$

If we have $QLQ^{-1} = L'$ with $\det Q$ relatively prime to $\det \Pi$, then under certain supplementary conditions imposed on $\det \Pi$, we have

$$\text{(VII.3.12)} \qquad \bar{\Pi}^T Q \equiv W (\bmod \det \Pi)\,,$$

where W is a symmetric matrix. We remark that for a Q which satisfies condition (VII.3.12) we have

$$\text{(VII.3.13)} \qquad Q\Pi = \Pi^T Q'\,,$$

where Q' is an integral matrix.

Indeed, we have from (VII.3.12) $(\bar{\Pi}^T Q)^T = W + dU$, where U is an integral matrix and $d = \det \Pi$. Further,

$$Q\Pi = \Pi^T (\bar{\Pi}^T Q) \bar{\Pi}^{-1} = \Pi^T (\bar{\Pi}^T Q) \Pi\, d^{-1}$$

and

$$(Q\Pi)^T = \Pi^T (\bar{\Pi}^T Q + dU_1) \Pi\, d^{-1} = \Pi^T (\bar{\Pi}^T Q) \Pi + \Pi^T U_2\,,$$

where U_1 and U_2 are integral matrices. Since $(Q\Pi)^T = \Pi^T Q^T$, (VII.3.13) follows.

In the case of matrices of higher order equation (VII.3.12) can play the role of equation (VII.3.10) in the case of 2×2-matrices (or quaternions). We remark yet that if the equation

$$l + L = UV$$

is valid, where U and V are symmetric matrices, then it follows from this that

$$l + L^T = V^T U^T = VU\,; \quad VLV^{-1} = L^T\,,$$

that is, V rotates the solution L of equation (VII.1.1) into the transposed solution L^T (which must then be a solution). The study of rotations of this kind is of independent interest. We formulate one remark concerning this problem (see LINNIK [7], p. 23).

We call the pair of matrices A, B a pair of weakly symmetric matrices if there exists a unimodular matrix ε such that $B = B^T\varepsilon$, $A = \varepsilon^{-1}A^T$, $BA = B^T A^T$, and $AB = A^T B^T$.

Let k be an algebraic field of degree n and let a be an integer free of squares and not dividing the discriminant d of the field.

Let θ be an integral element of the field having no rationally integral divisors and let $L_1, L_2, \ldots, L_h$ (where h is the class number of the field k) be the matrices corresponding to θ in all non-equivalent representations of the ring of integral elements of the field k by matrices. If the number a possesses certain properties (in relation to the degrees of our prime ideals), then the number of equations of the type

$$lE + L_i = A_i B_i, \quad \det A_i = a\,,$$

where A_i, B_i is a pair of weak symmetric matrices, is comparatively simply connected with the number of ideal divisors of a and the class number of second order of the class group H in the field k.

For quadratic fields $k(\sqrt{\pm D})$ the similar theorems are equivalent to a known theorem of Gauss on ternary representations of binary forms; our theorems turn out to be generalizations of this theorem.

We remark that for development of a theory analogous to the theory obtained for 2×2-matrices, we still need a generalization of Lemma V.11.1 on the asymptotic distribution of 2×2-matrices with large determinants. Such a generalization has been given by the author and B. F. SKUBENKO for 3×3-matrices (see LINNIK-SKUBENKO) and generalized to matrices of any order by B. F. SKUBENKO (see SKUBENKO [2]). We present this result in the following chapter, for simplicity restricting ourselves to the case of 3×3-matrices.

Chapter VIII

Asymptotic Distribution of Integral 3×3 Matrices

§ 1. Formulation of the Problem

In this chapter we study a generalization of Lemma V.11.1 to 3×3-matrices. For matrices of arbitrary order a corresponding asymptotic result has been proved by B. F. SKUBENKO [2]. The suitable measure, which we denote by meas Ω, is Haar measure on the group of unimodular 3×3 matrices.

Let N be a sufficiently large counting number and p a prime number which does not divide N.

We consider the integral 3×3 matrix

$$X = (x_{ij}), \quad \det X = N$$

and corresponding to it the matrix

$$\tilde{X} = \left(\frac{x_{ij}}{N^{\frac{1}{3}}}\right) = (\tilde{x}_{ij}).$$

We denote by the symbol Ω any integral domain (in the sense of Jordan) of unimodular 3×3 matrices, and by meas Ω the measure of this domain.

We consider further integral 3×3 matrices

$$A = (a_{ij}), \quad \det A \equiv N (\operatorname{mod} p).$$

Finally we are interested in a system of integral matrices X satisfying the conditions

(*) $$\tilde{X} \in \Omega, \ \det X = N, \ X \equiv A (\operatorname{mod} p).$$

The following theorem is valid.

Theorem VIII.1.1: *For the number of integral solutions of the system* (*) *as* $N \to \infty$ *we have asymptotically*

$$\Phi(\Omega, N, A, p) \sim \frac{\operatorname{meas} \Omega}{\zeta(2)\,\zeta(3)} N^2 \sum_{d \mid N} \frac{\sigma(d)}{d^2} f(p),$$

where p and Ω are fixed; $\sigma(d)$ is the sum of the divisors of d; and

$$f(n) = \begin{cases} 1 & \text{for } n = 1, \\ \dfrac{1}{p^3(p^3-1)(p^2-1)} & \text{for } n = p \text{ prime}. \end{cases}$$

Let two numbers N and p be given: N is a counting number and p is a prime not dividing N.

We consider the integral 3×3 matrix

(VIII.1.1) $$X = (x_{ij})$$

and the 3×3 matrix

(VIII.1.2) $$A = (a_{ij})$$

under the condition $0 \leqq a_{ij} < p$.

Lemma VIII.1.1: *The system*

(VIII.1.3) $$\begin{cases} \det X = N \\ X \equiv A \pmod p \end{cases}$$

has a solution under the following conditions:

1) $\det A \equiv 0 \pmod p$;

2) $a_{i_1 j_1}$ *is determined by* N *and the other components of the matrix* A, $a_{i_1 j_1}$ *being that element of the matrix* A *for which the algebraic complement* $A_{i_1 j_1} \not\equiv 0 \pmod p$;

3) $(N, p) = 1$.

Proof: Without loss of generality we may assume

(VIII.1.3′) $$A_{i_1 j_1} = A_{33} = \begin{pmatrix} a_{11} & a_{12} \\ a_{21} & a_{22} \end{pmatrix} \not\equiv 0 \pmod p .$$

First of all it is clear that

(VIII.1.4) $$\det X = N$$

is solvable. Let this solution be of the form

(VIII.1.5) $$\begin{vmatrix} x_{11} & x_{12} & x_{13} \\ x_{21} & x_{22} & x_{23} \\ x_{31} & x_{32} & x_{33} \end{vmatrix} = N$$

with the condition

$$\begin{pmatrix} x_{11} & x_{12} \\ x_{21} & x_{22} \end{pmatrix} \equiv \begin{pmatrix} a_{11} & a_{12} \\ a_{21} & a_{22} \end{pmatrix} \pmod p$$

(this is obviously possible). We set

$$q = \begin{vmatrix} x_{11} & x_{12} \\ x_{21} & x_{22} \end{vmatrix}, \quad X_{32} = \begin{vmatrix} x_{11} & x_{13} \\ x_{21} & x_{23} \end{vmatrix}, \quad X_{31} = \begin{vmatrix} x_{12} & x_{13} \\ x_{22} & x_{23} \end{vmatrix}.$$

Remark 1. Obviously, for any integral t_1, t_2, t_3, t_4 we can find a t such that

(VIII.1.5′) $$\begin{vmatrix} x_{11} & x_{12} & x_{13} + q t_1 \\ x_{21} & x_{22} & x_{23} + q t_2 \\ x_{31} + q t_3 & x_{32} + q t_4 & t \end{vmatrix} = N .$$

Therefore, in connection with $(q, p) = 1$, our proof of Lemma VIII.1.1 is completed.

Lemma VIII.1.2: *Lemma VIII.1.1 is true under the supplementary condition*

(VIII.1.6) $$(q, X_{32}, X_{31}) = 1 .$$

This is obvious.

Lemma VIII.1.3: *Suppose that the system* (VIII.1.3) *with condition* (VIII.1.6) *possesses the solution*

$$\begin{vmatrix} x_{11} & x_{12} & x_{13} \\ x_{21} & x_{22} & x_{23} \\ x_{31} & x_{32} & x_{33} \end{vmatrix} = N,$$

(VIII.1.7)

$$\begin{pmatrix} x_{11} & x_{12} & x_{13} \\ x_{21} & x_{22} & x_{23} \\ x_{31} & x_{32} & x_{33} \end{pmatrix} \equiv \begin{pmatrix} a_{11} & a_{12} & a_{13} \\ a_{21} & a_{22} & a_{23} \\ a_{31} & a_{32} & a_{33} \end{pmatrix} \pmod{p}.$$

Then each solution of the system (VIII.1.3) *with fixed* $\begin{pmatrix} x_{11} & x_{12} \\ x_{21} & x_{22} \end{pmatrix}$ *is of the form*

(VIII.1.8) $$X = \begin{pmatrix} x_{11} & x_{12} & zx_{13} + xx_{11} + yx_{12} + t_1 pq \\ x_{21} & x_{22} & zx_{33} + xx_{21} + yx_{22} + t_2 pq \\ z'x_{31} + \bar{x}x_{11} + \bar{y}x_{21} + t_3 pq & z'x_{32} + \bar{x}x_{12} + \bar{y}x_{22} + t_4 pq & t \end{pmatrix}$$

where $zz' \equiv 1 \pmod{qp}$, $z \equiv 1 \pmod{p}$, $x \equiv y \equiv \bar{x} \equiv \bar{y} \equiv 0 \pmod{p}$, *and where* t *is uniquely determined by* N *and the remaining components of the matrix* X.

Proof: It is obvious that (VIII.1.8) is a solution of the system (VIII.1.3). We show that (VIII.1.8) represents all solutions of the system (VIII.1.3). We consider an arbitrary solution of the system (VIII.1.3):

(VIII.1.9) $$X_1 = \begin{pmatrix} x_{11} & x_{12} & a_1 \\ x_{21} & x_{22} & a_2 \\ b_1 & b_2 & T \end{pmatrix}.$$

Using (VIII.1.6) we find x_1, y_1, z_1 with $x_1 \equiv y_1 \equiv 0 \pmod{p}$, $(z_1, qp) = 1$, such that

(VIII.1.10) $$\begin{cases} a_1 \equiv zx_{13} + x_1 x_{11} + y_1 x_{12} \pmod{pq}, \\ a_2 \equiv zx_{23} + x_1 x_{21} + y_1 x_{22} \pmod{pq}. \end{cases}$$

Then, because X_1 is a solution of the system (VIII.1.3), we can also find x_1, y_1 with $x_1 \equiv y_1 \equiv 0 \pmod{p}$, $z_1 \bar{z}_1 \equiv 1 \pmod{pq}$, such that

(VIII.1.11) $$\begin{cases} b_1 \equiv z_1' x_{31} + \bar{x}_1 x_{11} + \bar{y}_1 x_{21} \pmod{pq}, \\ b_2 \equiv z_1' x_{32} + \bar{x}_1 x_{12} + \bar{y}_1 x_{22} \pmod{pq}, \end{cases}$$

and Lemma VIII.1.3 is proved.

We remark that under a solution of the system (VIII.1.3) we shall understand a solution (VIII.1.8) in which $z, z', x, y, \bar{x}, \bar{y}$ are the smallest positive residues mod pq. Two solutions X_1 and X_2 will be called equal if and only if

$$X_1 \equiv X_2 \pmod{pq}.$$

Lemma VIII.1.4: *If $X^{(0)}$ is a solution of the system* (VIII.1.3), *then the number of solutions of this system with*

$$X \equiv X^{(0)} \pmod{pq} \tag{VIII.1.12}$$

and fixed $\begin{pmatrix} x_{11} & x_{12} \\ x_{21} & x_{22} \end{pmatrix}$ *in the conditions of Lemma VIII.1.3 is equal to q^2.*

Proof: Let

$$\begin{aligned} z_1 x_{13} + x_1 x_{11} + y_1 x_{12} &\equiv z_2 x_{13} + x_2 x_{11} + y_2 x_{12} \pmod{pq}, \\ z_1 x_{23} + x_1 x_{21} + y_1 x_{22} &\equiv z_2 x_{23} + x_2 x_{21} + y_2 x_{22} \pmod{pq}; \end{aligned} \tag{VIII.1.13}$$

then taking into account $z_1 - z_2 \equiv 0 \pmod{p}$ and

$$|z_1 - z_2| < pq, \quad \begin{vmatrix} x_{11} & x_{12} \\ x_{21} & x_{22} \end{vmatrix} = q, \; x_1 - x_2 \equiv y_1 - y_2 \pmod{p},$$

and also condition (VIII.1.6), we get

$$z_1 = z_2 . \tag{VIII.1.14}$$

Using (VIII.1.6) and (VIII.1.14) we obtain further that the number of solutions of the system (VIII.1.13) is q. Discussing the system

$$\begin{cases} z_1' x_{31} + \bar{x}_1 x_{11} + \bar{y}_1 x_{21} \equiv z_2' x_{31} + \bar{x}_2 x_{11} + \bar{y}_2 x_{21} \pmod{pq} \\ z_1' x_{32} + \bar{x}_1 x_{12} + \bar{y}_1 x_{22} \equiv z_2' x_{32} + \bar{x}_2 x_{12} + \bar{y}_2 x_{22} \pmod{pq} \end{cases} \tag{VIII.1.15}$$

in a similar manner (with the calculation $z_1' = z_2'$), we obtain Lemma VIII.1.4.

§ 2. Some Estimates

We introduce the following notation:

$$m = pq$$

$$\begin{cases} z\, x_{13} + x x_{11} + y x_{12} + t_1 m = V_1 \\ z\, x_{23} + x x_{21} + y x_{22} + t_2 m = V_2 \\ z' x_{31} + \bar{x} x_{11} + \bar{y} x_{21} + t_3 m = V_3 \\ z' x_{32} + \bar{x} x_{12} + \bar{y} x_{22} + t_4 m = V_4 \end{cases} \tag{VIII.2.1}$$

$$\omega = \left\{ \frac{a_i N^{\frac{1}{3}+\varepsilon_i}}{m} < \frac{V_i}{m} \leqq \frac{a_i' N^{\frac{1}{3}+\varepsilon_i'}}{m} \right\} \qquad (i = 1, 2, 3, 4,), \tag{VIII.2.2}$$

a_i and a_i' are arbitrary fixed numbers with

$$a_i' N^{\varepsilon_i'} - a_i N^{\varepsilon_i} > N^{-\frac{1}{150}}, \; |\varepsilon_i|, \; |\varepsilon_i'| \leqq \tfrac{1}{150} \qquad (i = 1, 2, 3, 4).$$

Lemma VIII.2.1: *The number of distinct solutions of the system*

$$\det X = N$$

$$X \equiv A \pmod{p}$$

for fixed $\begin{pmatrix} x_{11} & x_{12} \\ x_{21} & x_{22} \end{pmatrix}$ *with the conditions*

$$\begin{aligned} &|x_{11}| \asymp q^{\frac{1}{2}+\eta_1}, \; |x_{12}| \asymp q^{\frac{1}{2}+\eta_2}, \; |x_{21}| \asymp q^{\frac{1}{2}+\eta_3}, \\ &|x_{22}| \asymp q^{\frac{1}{2}+\eta_4}, \; |q| \asymp N^{\frac{2}{3}+\eta_0}, \\ &(X_{13}, X_{23}, N) < q^{\frac{1}{100}}, \end{aligned} \tag{VIII.2.3}$$

where $|\eta_i| < \frac{1}{100}$ ($i=0, 1, 2, 3, 4$) *and where the condition* (VIII.2.2) *in the conditions of Lemma VIII.1.3 is fulfilled for* V_i ($i=1, 2, 3, 4$), *is*

$$R(\omega, q, A, N, \eta_i) = \operatorname{meas}\omega \frac{N^{4/3}\varphi(q)}{q^2 p^4}\,(1+0(N^{-\frac{1}{75}})),$$

where

$$\operatorname{meas}\omega = \prod_{i=1}^{4} (a_i' N^{\varepsilon_i'} - a_i N^{\varepsilon_i}).$$

Proof: We set

(VIII.2.4) $$\Delta_i = \frac{a_i N^{\frac{1}{3}+\varepsilon_i}}{m}, \quad \Delta_i' = \frac{a_i' N^{\frac{1}{3}+\varepsilon_i'}}{m} \qquad (i=1,2,3,4).$$

Without loss of generality we may assume

(VIII.2.5) $$[\Delta_i] = [\Delta_i'] \qquad (i=1,2,3,4).$$

It is necessary for us to compute the number of solutions of (VIII.2.3) with the property

$$\Delta_i < \{V_i/m\} \leqq \Delta_i' \qquad (i=1,2,3,4).$$

Here we apply the lemma on the expansion of the function

$$\Psi(V_1/m, V_2/m, V_3/m, V_4/m) = \Psi(V/m)$$

in a Fourier series with period 1.

We set

1) $\Psi(V/m)=1$ for $\Delta_i < \{V_i/m\} \leqq \Delta_i'$, $(i=1,2,3,4)$;
2) $0 \leqq \Psi(V/m) \leqq 1$ for

(VIII.2.6) $$\begin{cases} -(\Delta_i' - \Delta_i) m^{-\frac{1}{50}} + \Delta_i < \{V_i/m\} \leqq \Delta_i; \\ \Delta_i' < \{V_i/m\} \leqq \Delta_i' + (\Delta_i' - \Delta_i) m^{-\frac{1}{50}}; \\ \text{if for only one index} \\ -(\Delta_i' - \Delta_i) m^{-\frac{1}{50}} + \Delta_i < \{V_i/m\} \leqq \\ \leqq \Delta_i' + (\Delta_i' - \Delta_i) m^{-\frac{1}{50}} \end{cases} \qquad (i=1,2,3,4);$$

3) $\Psi(V/m)=0$ in the other cases.

Making use of Lemma VIII.1.4, we find that

(VIII.2.7) $$q^2 R(\omega, q, A, p, N, \eta_i) = \sum_{V_i (\operatorname{mod} m)} \Psi(V/m) + 0\left(q^2 \operatorname{meas}\omega \frac{N^{3/4}\varphi(q)}{q^2} m^{-\frac{1}{50}}\right),$$

(VIII.2.8) $$\Psi\left(\frac{V}{m}\right) = \sum_{c_1,c_2,c_3,c_4=-\infty}^{+\infty} D_{c_1c_2c_3c_4} \exp\left(2\pi i \frac{V_1c_1+V_2c_2+V_3c_3+V_4c_4}{m}\right).$$

On the coefficients we impose the bounds:

(VIII.2.8′) $$\begin{gathered} 1)\ |D_{c_1c_2c_3c_4}| \leqq \prod_{i=1}^{4} (\Delta_i' - \Delta_i), \quad |D_{c_1c_2c_3c_4}| \leqq \frac{1}{(\lambda_1\lambda_2\lambda_3\lambda_4)^r} \\ 2)\ \lambda_i \geqq \max\left(\left(\frac{1}{\Delta_i' - \Delta_i}\right)^{\frac{1}{r}}, \frac{1}{r}\left| m^{-\frac{1}{50}} (\Delta_i' - \Delta_i) c_i \right|\right), \end{gathered}$$

r being an arbitrary counting number.

We substitute in $\sum_{V_i (\operatorname{mod} m)} \Psi(V/m)$ for V_i its value from (VIII.2.1). We have

$$\sum_{\substack{V_j (\operatorname{mod} m)\\ (j=1,2,3,4)}} \Psi\left(\frac{V}{m}\right) = \sum_{V_j (\operatorname{mod} m)} \sum_{c_1,c_2,c_3,c_4=-\infty}^{+\infty} D_{c_1c_2c_3c_4} \exp\left(2\pi i \frac{\sum_{j=1}^{4} c_j V_j}{m}\right) = \sum_{c_1,c_2,c_3,c_4=-\infty}^{+\infty} D_{c_1c_2c_3c_4} \sum_{V_j (\operatorname{mod} m)} \exp\left(2\pi i \frac{\sum_{j=1}^{4} c_j V_j}{m}\right)$$

$$= \sum_{c_1,c_2,c_3,c_4=-\infty}^{+\infty} D_{c_1c_2c_3c_4} \sum_{\left.\begin{array}{r} zz' \equiv 1 (\operatorname{mod} m)\\ z \equiv 1 (\operatorname{mod} m)\\ x \equiv y \equiv \bar{x} \equiv \bar{y} \equiv 0 (\operatorname{mod} p)\end{array}\right\} (\operatorname{mod} m)} \exp\left(2\pi i \frac{\begin{vmatrix} c_2 & x_{13}\\ -c_1 & x_{23}\end{vmatrix} z + \begin{vmatrix} x_{11} & -c_2\\ x_{21} & c_1\end{vmatrix} x + \begin{vmatrix} c_2 & x_{12}\\ -c_1 & x_{22}\end{vmatrix} y + \begin{vmatrix} c_4 & -c_3\\ x_{31} & x_{32}\end{vmatrix} z^1 + \begin{vmatrix} x_{11} & x_{12}\\ -c_4 & c_3\end{vmatrix} \bar{x} + \begin{vmatrix} c_4 & -c_3\\ x_{21} & x_{22}\end{vmatrix} \bar{y}}{m}\right)$$

$$= \sum_{c_1,c_2,c_3,c_4=-\infty}^{+\infty} D_{c_1c_2c_3c_4} \sum_{\substack{z=1\\ z' \equiv 1 (\operatorname{mod} p)\\ zz' \equiv 1 (\operatorname{mod} m)}}^{m-1} \exp\left(2\pi i \frac{\begin{vmatrix} c_2 & x_{13}\\ -c_1 & x_{23}\end{vmatrix} z + \begin{vmatrix} c_4 & -c_3\\ x_{31} & x_{32}\end{vmatrix} z'}{m}\right) \sum_{x=0}^{q-1} \exp\left(2\pi i \frac{\begin{vmatrix} x_{11} & -c_2\\ x_{21} & c_1\end{vmatrix} x}{q}\right) \sum_{y=0}^{q-1} \exp\left(2\pi i \frac{\begin{vmatrix} c_2 & x_{12}\\ -c_1 & x_{22}\end{vmatrix} y}{q}\right) \times$$

$$\times \sum_{\bar{x}=0}^{q-1} \exp\left(2\pi i \frac{\begin{vmatrix} x_{11} & x_{12}\\ -c_4 & c_3\end{vmatrix} \bar{x}}{q}\right) \sum_{\bar{y}=0}^{q-1} \exp\left(2\pi i \frac{\begin{vmatrix} c_4 & -c_3\\ x_{21} & x_{22}\end{vmatrix} \bar{y}}{q}\right)$$

$$= \prod_{j=1}^{4} (\Delta_j' - \Delta_j) q^4 \sum_{\substack{z=1\\ z \equiv 1 (\operatorname{mod} p)}}^{m-1} 1 + \sum_{\substack{c_1,c_2,c_3,c_4=-\infty\\ |c_1|+|c_2|+|c_3|+|c_4| \neq 0}}^{+\infty} D_{c_1c_2c_3c_4} \sum_{V_j (\operatorname{mod} m)} \exp\left(2\pi i \frac{\sum_{j=1}^{4} c_j V_j}{m}\right).$$

Obviously,

(VIII.2.9)
$$\sum_{\substack{z=1 \\ z\equiv 1 \pmod p}}^{m-1} 1 = \varphi(m/p) + 0(m^{\xi}) = \varphi(q) + 0(q^{\xi})$$

(ξ as small as desired).

We estimate from above

$$\left| \sum_{\substack{c_1,c_2,c_3,c_4=-\infty \\ |c_1|+|c_2|+|c_3|+|c_4|\neq 0}}^{+\infty} D_{c_1c_2c_3c_4} \sum_{V_i \pmod m} \exp\left(2\pi i \frac{\sum_{j=1}^{4} c_j V_j}{m}\right) \right| = M$$

We make use of the known estimate (see A. WEIL)

$$\left| \sum_{\substack{z=1 \\ zz'\equiv 1 \pmod m \\ z\equiv 1 \pmod p \\ p|m}}^{m-1} \exp\left(2\pi i \frac{k_1 z + k_2 z'}{m}\right) \right| < m^{\frac{1}{2}+\xi} \min\left(\sqrt{(m,k_1)},\ \sqrt{(m,k_2)}\right)$$

(ξ as small as desired).

Then,

$$M \leq \left| \sum_{\substack{c_1,c_2,c_3,c_4=-\infty \\ |c_1|+|c_2|+|c_3|+|c_4|\neq 0 \\ |c_j| \leq m^{\varepsilon+\frac{1}{50}} \frac{1}{\Delta_j'-\Delta_j}}}^{+\infty} D_{c_1c_2c_3c_4} \sum_{V_j \pmod m} \exp\left(2\pi i \frac{\sum_{j=1}^{4} c_j V_j}{m}\right) \right| +$$

$$+ \left| \sum_{\substack{c_1,c_2,c_3,c_4=-\infty \\ |c_1|+|c_2|+|c_3|+|c_4|\neq 0 \\ \max_{j=1,2,3,4} |c_j| > m^{\varepsilon+\frac{1}{50}} \frac{1}{\Delta_j'-\Delta_j}}}^{+\infty} D_{c_1c_2c_3c_4} \sum_{V_j \pmod m} \exp\left(2\pi i \frac{\sum_{j=1}^{4} c_j V_j}{m}\right) \right| .$$

The latter summand in connection with (VIII.2.8′) for sufficiently large m and for a fixed ε as small as desired can be estimated by $0(q^2)$ (this will suffice for our purposes).

We estimate the first summand. For this we need an upper estimate for the number of solutions to the congruences

$$\begin{vmatrix} x_{11} & -c_2 \\ x_{21} & c_1 \end{vmatrix} \equiv \begin{vmatrix} c_2 & x_{12} \\ -c_1 & x_{22} \end{vmatrix} \equiv \begin{vmatrix} x_{11} & x_{12} \\ -c_4 & c_3 \end{vmatrix} \equiv \begin{vmatrix} c_4 & -c_3 \\ x_{21} & x_{22} \end{vmatrix} \equiv 0 \pmod q,$$

(VIII.2.10)
$$|c_j| \leq m^{\varepsilon+\frac{1}{50}} \frac{1}{\Delta_j'-\Delta_j} \qquad (j=1,2,3,4).$$

We set

$$d = \min\left(\left(q, \begin{vmatrix} c_2 & x_{13} \\ -c_1 & x_{23} \end{vmatrix}\right),\ \left(q, \begin{vmatrix} c_4 & -c_3 \\ x_{31} & x_{32} \end{vmatrix}\right)\right)$$

and estimate for this d the number of solutions to the congruences (VIII.2.10) by a calculation using the condition of Lemma VIII.2.1; $c_2 \equiv c_1 \equiv 0 \pmod{d'}$ follows from (VIII.1.6), where $d' \geq d$.

Further, since

$$\begin{vmatrix} x_{11} & -c_2 \\ x_{21} & c_1 \end{vmatrix}, \begin{vmatrix} c_2 & x_{12} \\ -c_1 & x_{22} \end{vmatrix}, \begin{vmatrix} x_{11} & x_{12} \\ -c_4 & c_3 \end{vmatrix}, \begin{vmatrix} c_4 & -c_3 \\ x_{21} & x_{22} \end{vmatrix} \leqq$$
$$\leqq \text{const} \cdot q^{\frac{1}{2}+\frac{1}{100}} m^{\varepsilon+\frac{1}{50}} \max_{j=1,2,3,4} (qN^{-\frac{1}{3}-\varepsilon_j}, qN^{-\frac{1}{3}-\varepsilon_j'})$$

and since

$$\begin{vmatrix} x_{11} & -c_2 \\ x_{21} & c_1 \end{vmatrix} = qt_1, \quad \begin{vmatrix} c_2 & x_{12} \\ -c_1 & x_{22} \end{vmatrix} = qt_2, \quad \begin{vmatrix} x_{11} & x_{12} \\ -c_4 & c_3 \end{vmatrix} = qt_3,$$
$$\begin{vmatrix} c_4 & -c_3 \\ x_{21} & x_{22} \end{vmatrix} = qt_4,$$

then the number of solutions is of the order

(VIII.2.11) $$0(q^{\frac{1}{20}+\frac{1}{100}} q^{\frac{1}{20}}).$$

Hence it follows that the first summand, and at the same time M, is estimated by

$$M \ll \prod_{i=1}^{4} (\Delta_i' - \Delta_i) q^4 q^{\frac{1}{2}+\xi} q^{\frac{1}{200}} q^{\frac{1}{20}+\frac{1}{100}} q^{\frac{1}{20}} + q^2 \ll \prod_{i=1}^{4} (\Delta_i' - \Delta_i) q^4 q^{1-\frac{2}{5}}.$$

Taking into account

$$\prod_{i=1}^{4} (\Delta_i' - \Delta_i) = \prod_{i=1}^{4} (a_i' N^{\varepsilon_i'} - a_i N^{\varepsilon_i}) \frac{N^{3/4}}{q^4 p^4}$$

and formulas (VIII.2.7), (VIII.2.9), we obtain the proof to Lemma VIII.2.1.

§ 3. Completion of the Proof

We consider the matrix

(VIII.3.1) $$Y = \begin{pmatrix} x_{11} & x_{12} \\ x_{21} & x_{22} \end{pmatrix}$$

and we solve the system

$$\det Y = q$$

(VIII.3.2) $$Y \equiv \begin{pmatrix} a_{11} & a_{12} \\ a_{21} & a_{22} \end{pmatrix} \pmod p$$

under the conditions

1) $$b_1 q^{\kappa_1} < x_{11}/q^{1/2} \leqq b_1' q^{\kappa_1'}, \; b_2 q^{\kappa_2} < x_{21}/q^{1/2} \leqq b_2' q^{\kappa_2'},$$

(VIII.3.3) $$b_3 q^{\kappa_3} < x_{12}/x_{11} \leqq b_3' q^{\kappa_3'},$$

where $b_1, b_1', b_2, b_2', b_3, b_3'$ are fixed numbers, $b_i' - b_i > 0$, $|\kappa_i|, |\kappa_i'| \leqq \frac{1}{100}$ $(i = 1, 2, 3)$;

2) $$(x_{11}, x_{12}, x_{21}, x_{22}) = 1.$$

We formulate the following lemma (its proof was provided in Chapter V, § 11 — Lemma V.11.1).

Lemma VIII.3.1: *The number of solutions W of the system* (VIII.3.2) *under the conditions* 1), 2) *is*

$$W\left(p\begin{pmatrix} a_{11} & a_{12} \\ a_{21} & a_{22}\end{pmatrix}, q\right) = \sum_{d^2|q} \mu(d) \sum_{r|\frac{q}{d^2}} \frac{q}{d^2 r} \frac{\operatorname{meas}\omega_1}{\zeta(2)p(p^2-1)} (1 + 0(q^{-\frac{1}{200}})),$$

where

$$\operatorname{meas}\omega_1 = \prod_{i=1}^{3} (b'_i q^{\kappa'_i} - b_i q^{\kappa_i}), \quad \zeta(2) = \sum_{n=1}^{\infty} \frac{1}{n^2}.$$

Remark. In order to apply this lemma for proving our theorem, we call to mind that we imposed conditions on $x_{11}, x_{12}, x_{21}, x_{22}$ in Lemma VIII.2.1 which are satisfied in the lemma formulated above if q is of the order $N^{\frac{2}{3}+\eta_0}$, $|\eta_0| < \frac{1}{100}$. Further we note that the conditions of Lemma VIII.2.1 require $(N, x_{11}, x_{21}) = d < q^{\frac{1}{100}}$. With this requirement the lemma just formulated is valid (this is easily seen by examining the proof).

Lemma VIII.3.2: *The number of solutions of the system*

$$\det X = N$$
$$X \equiv A \pmod p$$

under the conditions of Lemmas VIII.2.1 and VIII.3.1 is

(VIII.3.4)
$$f(N, \omega_1, \omega, A, p) = \frac{\operatorname{meas}\omega_1 \ \operatorname{meas}\omega}{\zeta(2)p^5(p^2-1)} N^{4/3} \sum_{d^2|q} \mu(d) \times \sum_{r|\frac{q}{d^2}} \frac{1}{d^2 r} \frac{\varphi(q)}{q} (1 + 0(N^{-\frac{1}{199}})).$$

Proof: This follows from (VIII.3.3′) because in this case the system is always solvable.

Lemma VIII.3.3: *The number of the solutions of the system*

$$\det X = N$$
$$X \equiv A \pmod p$$

under the conditions of Lemma VIII.3.2 and the supplementary requirement that q take values in the interval

(VIII.3.5)
$$\{N^{\frac{2}{3}+\eta_0} c_1 < q \leqq N^{\frac{2}{3}+\eta'_0} c'_i\}$$

is

(VIII.3.6)
$$F(N, \omega, \omega_1, \omega_2, A, p) = \frac{\operatorname{meas}\omega \operatorname{meas}\omega_1 \operatorname{meas}\omega_2}{\zeta(2)\,\zeta(3)\,p^3(p^3-1)\,(p^2-1)} N^2 (1 + 0(N^{-\frac{1}{200}})).$$

Proof: We note first of all that

$$q \equiv \begin{vmatrix} a_{11} & a_{12} \\ a_{21} & a_{22}\end{vmatrix} \pmod p.$$

We put

$$a = \begin{vmatrix} a_{11} & a_{12} \\ a_{21} & a_{22}\end{vmatrix}.$$

Further, since

$$\sum_{\substack{N^{2/3+\eta_0 c_1} < q \leq N^{2/3+\eta_0 c'_1} \\ q \equiv a (\bmod p)}} \frac{\varphi(q)}{q} \sum_{d^2 | q} \mu(d) \sum_{r | \frac{q}{d^2}} \frac{1}{d^2 r} = \sum_{\substack{N^{2/3+\eta_0 c_1} < q \leq N^{2/3+\eta_0 c'_1} \\ q \equiv a (\bmod p)}} \sum_{d | q} \frac{\mu(d)}{d^2},$$

Lemma VIII.3.3 follows because of

$$\sum_{\substack{q \leq k \\ q \equiv a (\bmod p) \\ a \not\equiv 0 (\bmod p)}} \sum_{d | q} \frac{\mu(d)}{d^2} = \frac{k p^2}{\zeta(3)(p^3-1)} + 0(\log M).$$

We are now close to completion of the proof of our theorem. Lemma VIII.3.3 was proved subject to the condition (VIII.1.6). We want to deduce an analogous theorem without condition (VIII.1.6).

Let

(VIII.3.7) $$(q, X_{32}, X_{31}) = d = N_1 N_2.$$

Then we can put

(VIII.3.8) $$\det X = N = N' N_1 N_2.$$

Consequently there exists a matrix

(VIII.3.9) $$X'' = \begin{pmatrix} N_1 & 0 & 0 \\ \alpha & N_2 & 0 \\ 0 & 0 & 1 \end{pmatrix}, \quad 0 \leq \alpha < N_2,$$

such that

(VIII.3.10) $$X = X'' X', \quad \det X' = N',$$

and for the matrix X' the condition (VIII.1.6) is satisfied.

We set $A' \equiv (X'')^{-1} A \,(\bmod p)$ and solve the system

(VIII.3.11) $$\begin{gathered} \det X' = N' \\ X' \equiv A' (\bmod p). \end{gathered}$$

Further, we make the following replacements:

for (VIII.2.2),

(VIII.3.12) $$\begin{aligned} \omega' = \Big\{ & \frac{a_1 N^{\frac{1}{3}+\varepsilon_1}}{N_1} < V'_1 \leq \frac{a'_1 N^{\frac{1}{3}+\varepsilon'_1}}{N_1}, \; -\frac{\alpha a_1 N^{\frac{1}{3}+\varepsilon_1}}{N_1 N_2} + \\ & + \frac{a_2 N^{\frac{1}{3}+\varepsilon_2}}{N_2} < V'_2 \leq \frac{-\alpha a'_1 N^{\frac{1}{3}+\varepsilon'_1}}{N_1 N_2} + \frac{a'_2 N^{\frac{1}{3}+\varepsilon'_2}}{N_2}, \\ & a_i N^{\frac{1}{3}+\varepsilon_i} < V'_i \leq a'_i N^{\frac{1}{3}+\varepsilon'_i} \Big\} \quad (i = 3, 4); \end{aligned}$$

for (VIII.3.3),

(VIII.3.13) $$\begin{aligned} \omega'_1 = \Big\{ & \frac{b_1 q^{\frac{1}{2}+\kappa_1}}{N_1} < x'_{11} \leq \frac{b'_1 q^{\frac{1}{2}+\kappa'_1}}{N_1}, \; b_3 q^{\kappa_3} < \frac{x'_{12}}{x'_{11}} \leq b'_3 q^{\kappa'_3}, \\ & -\frac{b_1 q^{\frac{1}{2}+\kappa_1}}{N_1 N_2} + \frac{b_2 q^{\frac{1}{2}+\kappa_2}}{N_2} < x'_{21} \leq \frac{-\alpha b'_1 q^{\frac{1}{2}+\kappa'_1}}{N_1 N_2} + \frac{b_2 q^{\frac{1}{2}+\kappa'_2}}{N_2} \Big\}; \end{aligned}$$

for (VIII.3.5),

(VIII.3.14) $$\omega_2' = \left\{ \frac{N^{\frac{2}{3}+\eta_0}}{N_1 N_2} c_1 < q' \leqq \frac{N^{\frac{2}{3}+\eta_0} c_1'}{N_1 N_2} \right\}.$$

Let $N_1 N_2 \leqq N^{\frac{1}{1500}}$. We arrange the magnitudes of

$$\varepsilon_i (i=1,2,3,4)\,; \quad \kappa_i (i=1,2,3)\,; \quad \eta_i (i=1,2,3,4)\,; \quad \eta_0 .$$

We set

$$|\eta_0| \leqq \tfrac{1}{300}; \ |\eta_i|, |\eta_i'| \leqq \tfrac{1}{200}; \ |\varepsilon_i|, |\varepsilon_i'| \leqq \tfrac{1}{300}; \ |\kappa_i|, |\kappa_i'| \leqq \tfrac{1}{200};$$

(VIII.3.15) $$a_1' N^{\varepsilon_1'} - a_1 N^{\varepsilon_1} < (a_2' N^{\varepsilon_2'} - a_2 N^{\varepsilon_2}) N^{-\frac{1}{500}}\,;$$
$$b_1' q^{\kappa_1'} - b_1 q^{\kappa_1} < (b_2' q^{\kappa_2'} - b_2 q^{\kappa_2}) N^{-\frac{1}{500}}\,.$$

Then the following lemma is valid.

Lemma VIII.3.4: *The number of solutions of the system*

$$\det X' = N'$$
$$X' \equiv A' \pmod p$$

satisfying the conditions (VIII.3.12), (VIII.3.13), (VIII.3.14), (VIII.3.15) *is*

$$F\left(\frac{N}{N_1 N_2}, \omega', \omega_1', \omega_2', A', p\right) = \frac{N^2}{N_1^2 N_2^2} \frac{\operatorname{meas}\omega \operatorname{meas}\omega_1 \operatorname{meas}\omega_2}{\zeta(2)\zeta(3) p^3 (p^2-1)(p^3-1)} \times$$
$$\times (1 + 0(N^{-\frac{1}{1000}}))\,.$$

Proof: Taking into account (VIII.3.15), we find that

$$\operatorname{meas}\omega' = \frac{1}{N_1 N_2} \operatorname{meas}\omega \ (1 + 0(N^{-\frac{1}{500}}))\,,$$

$$\operatorname{meas}\omega_1' = \frac{1}{N_1 N_2} \operatorname{meas}\omega_1 (1 + 0(N^{-\frac{1}{500}}))\,,$$

$$\operatorname{meas}\omega_2' = \frac{1}{N_1 N_2} \operatorname{meas}\omega_2\,.$$

Further, using (VIII.3.13), we obtain from Lemma VIII.3.1

$$W'\left(p, \begin{pmatrix} a_{11} & a_{12} \\ a_{21} & a_{22} \end{pmatrix}, q'\right) = \sum_{d|q'} \mu(d) \sum_{r|\frac{q'}{d^2}} \frac{q'}{d^2 r} \frac{N_1 N_2 \operatorname{meas}\omega_1'}{\zeta(2) p(p^2-1)} \times$$
$$\times (1 + 0(q^{-\frac{1}{300}}))\,.$$

After these remarks Lemma VIII.3.4 is obvious.

We denote by Ω_1 the intersection of the regions ω, ω_1, ω_2:

$$\operatorname{meas}\Omega_1 = \operatorname{meas}\omega \operatorname{meas}\omega_1 \operatorname{meas}\omega_2$$
$$= \prod_{i=1}^{4} (a_i' N^{\varepsilon_i'} - a_i N^{\varepsilon_i}) \prod_{i=1}^{3} (b_i' q^{\kappa_i'} - b_i q^{\kappa_i}) (c_1' - c_1) N^{\eta_0}\,.$$

Lemma VIII.3.5: *The number of integral solutions of the system*

$$\det X = N$$
$$X \equiv A \pmod p$$

with $X \in \Omega_1$, $(N,p)=1$, *and* $\det A \not\equiv 0 \pmod p$ *is*

$$\Phi(\Omega_1, N, X\in\Omega_1, X\equiv A \pmod p) = \sum_{d|N}\left(\frac{N}{d}\right)^2 \sum_{r|d}\frac{d}{r}\,\frac{\operatorname{meas}\Omega_1}{\zeta(2)\,\zeta(3)}\times$$
$$\times\frac{1}{p^3(p^3-1)(p^2-1)}\left(1+0(N^{-\frac{1}{2000}})\right).$$

Proof: We turn to the relation (VIII.3.10):

$$X = X''X' = \begin{pmatrix} N_1 & 0 & 0\\ \alpha & N_2 & 0\\ 0 & 0 & 1\end{pmatrix}\cdot X'.$$

It is clear that the number of such X for which $\det X'' = N_1 N_2$ will possess a principal term of the form

$$\sum_{d|N_1N_2}\frac{N_1N_2}{d}\,F\left(\frac{N}{N_1N_2}, \omega', \omega_1', \omega_2', A', p\right).$$

Hence Lemma VIII.3.5 is proved, since $X\in\Omega_1$. Our theorem follows immediately from Lemma VIII.3.5 Indeed, for η_i, ε_i, $\kappa_i<0$ with the conditions (VIII.3.15), and making use of

$$\operatorname{meas}\Omega = \int_\Omega \frac{d\tilde{x}_{11}\ldots d\tilde{x}_{13}\ldots d\tilde{x}_{31}\,d\tilde{x}_{32}}{\begin{vmatrix}\tilde{x}_{11} & \tilde{x}_{12}\\ \tilde{x}_{21} & \tilde{x}_{22}\end{vmatrix}},$$

we obtain $\operatorname{meas}\Omega_1 = \delta\,(\operatorname{meas}\Omega^{(j)})$; this means that

(VIII.3.16) $$\operatorname{meas}\Omega = \sum_{\substack{\Omega^{(j)}\in\Omega\\ \Omega^{(j)}\cap\Omega^{(f)}=\emptyset \text{ if } j\neq f}} \delta(\operatorname{meas}\Omega^{(j)}) + o(1).$$

The sum is taken over all $\Omega^{(j)}$.

With relation (VIII.3.16) as a foundation and setting

$$\delta\tilde{x}_{11}^{(j)} = b_1' N^{\kappa_1'} - b_1 N^{\kappa_1};\quad \frac{\delta(\tilde{x}_{12}^{(j)})}{\tilde{x}_{11}^{(j)}} = N^{\kappa_2'}b_2' - N^{\kappa_2}b_2;$$
$$\vdots \qquad\qquad\qquad\qquad \vdots$$
$$\delta\tilde{x}_{31}^{(j)} = a_3' N^{\varepsilon_3'} - a_3 N^{\varepsilon_3};\quad \delta(\tilde{x}_{32}^{(j)}) = a_4' N^{\varepsilon_4'} - a_4 N^{\varepsilon_4},$$

we have

$$\sum_\Omega \delta(\operatorname{meas}\Omega^{(j)})$$
$$= \sum_\Omega \delta\tilde{x}_{11}^{(j)}\,\delta\tilde{x}_{12}^{(j)}\,\frac{1}{\tilde{x}_{11}^{(j)}}\,\delta\tilde{x}_{21}^{(j)}\,\delta(\tilde{q}^{(j)})\,\frac{1}{\tilde{q}^{(j)}}\,\delta\tilde{x}_{13}^{(j)}\,\delta\tilde{x}_{23}^{(j)}\,\delta\tilde{x}_{31}^{(j)}\,\delta\tilde{x}_{32}^{(j)}$$
$$= \sum_\Omega \frac{\delta\tilde{x}_{11}^{(j)}\ldots\delta\tilde{x}_{13}^{(j)}\ldots\delta\tilde{x}_{31}^{(j)}\,\delta\tilde{x}_{32}^{(j)}}{\begin{vmatrix}\tilde{x}_{11}^{(j)} & \tilde{x}_{12}^{(j)}\\ \tilde{x}_{21}^{(j)} & \tilde{x}_{22}^{(j)}\end{vmatrix}}.$$

Therefore (VIII.3.16) will be satisfied.

Chapter IX

Further Generalizations. A Connection with the Generalized Riemann Hypothesis

§ 1. Further Generalizations

Up until now we considered representations of fields of n-th degree over the field of rational numbers by $n \times n$ matrices with integral coefficients or (in the case of quaternions) by special matrices with order two times the order of the field. However, we could take as a basic field not the field of rationals but any other algebraic number field and use matrices with coefficients belonging to the said field. In this vein, in an equation of the form (VII.1.1)

$$L^n + a_1(D) L^{n-1} + \cdots + a_n(D) = 0 \tag{IX.1.1}$$

which determines the field, we can assume that the numbers $a_j(D)$ $(j = 1, 2, \ldots, n)$ belong to a certain field k_0, fixed for all D; the coefficients of the matrices L must also belong to k_0. If the field k_0 is of degree a_0, then the system of equations (VIII.1.5) is changed into a system of equations in integers. This diophantine system will have $a_0^2 n^2$ variables and will consist of $a_0 n$ equations, and the asymptoticity of its solutions for $D \to \infty$ will also be inaccessible by the usual methods of analytic number theory, while the methods studied above may lead to non-trivial results. In particular, we take $n = 2$ and consider the basic equation

$$L^2 = -m, \tag{IX.1.2}$$

where m belongs to the field k_0 of degree a_0, k_0 being totally real (real together with all of the fields conjugate to it). Let L be a quaternion with coefficients belonging to the ring of integral elements of the field k_0. Then equation (IX.1.1) is equivalent to a problem concerning the asymptoticity of representations of a_0 numbers with special types of quadratic forms in $3a_0$ variables. This problem is inaccessible by the usual methods for any values $1, 2, \ldots,$ of a_0, including methods connected with the theory of modular functions.

By way of illustration we take the special case where the field k_0 is $k_0 = k\sqrt{3}$; $m = m_1 + m_2\sqrt{3}$; with m_1 and m_2 rational integers and m_2 even. We get a system of two equations of the form

$$\begin{aligned} x_1^2 + x_2^2 + x_3^2 + 3y_1^2 + 3y_2^2 + 3y_3^2 &= m_1 \\ x_1 y_1 + x_2 y_2 + x_3 y_3 &= m_2/2, \end{aligned} \tag{IX.1.3}$$

where x_i, y_i are rational integers. It is natural to require that $m_1 \to \infty$, m_2 being of the order of m_1; we see then that $x_j = 0(m_1^{1/2})$, $y_j = 0(m_1^{1/2})$ $(j = 1, 2, 3)$. In particular the case $m_2 = 0$ and the asymptotic geometry of solutions of this case are interesting.

§ 2. A Connection with the Generalized Riemann Hypothesis and its Weaker Forms

For the problems treated in Chapters IV, V, and VI (quaternions and 2×2 matrices) we also considered assertions which could be made if we accepted the Riemann hypothesis or one of its weaker forms. Such assertions are found in LINNIK [6]; however, sufficiently detailed research in this direction has only been carried out by A. V. MALYSHEV (pp. 179–190 of the reference) for the case of integral points on ellipsoids.

In Chapters IV, V, and VI we introduced an auxiliary prime number p (in Chapter IV it was denoted by q) such that $(-m/p) = 1$ or $(\pm D/p) = 1$. For ergodic theorems and mixing theorems its presence was natural, since the flow of primitive points was determined by it and the number entered into the corresponding formulations. But for the theorems on the distributions of integral points on regions of the second order surfaces we studied the presence of such a number turned out to be only a deficiency of the method and not an essential part of the problem.

We have seen that certain (unproved) hypotheses concerning the zeros of corresponding Dirichlet series would permit removal of this deficiency. Thus, for example, an improvement in the formulation of Theorem IV.1.1 follows from the cited investigations of A. V. MALYSHEV.

We introduce the Dirichlet L-series $L(s, X)$ with character $X(n) = (4m/n)$ modulo $4m$; $s = \sigma + ti$. Consider for sufficiently large m the region

$$(IX.2.1) \qquad |s - 1| < \frac{(\ln \ln m)^2 \ln \ln \ln m}{\sqrt{\ln m}}; \quad \sigma > 1 .$$

If there are no zeros of the series $L(s, X)$ in this region, then in Theorem IV.1.1 the remainder terms $\kappa_0(\Gamma_0, m)$ and $\kappa(\Gamma_0, m)$ are not dependent on the parameters and tend to zero for fixed Γ_0 as $m \to \infty$. The analogous statements may be made for the respective theorems of Chapters V and VI.

§ 3. Elementary Ergodic Considerations

In BREDIHIN-LINNIK new methods of approach to the study of the asymptotic geometry of solutions of certain diophantine equations are given with the help of the sieve of Eratosthenes and elementary ergodic considerations. Here the objects of investigation are equations of the form

$$(IX.3.1) \qquad a_l + \varphi(\xi, \eta) = m ,$$

where $\{a_l\}$ is some sequence of counting numbers and φ is a positive quadratic form with fixed determinant d; the number m approaches infinity.

It is assumed that the asymptoticity of the solutions to the auxiliary equation

$$a_l + \varrho(\xi, \eta) = m \tag{IX.3.2}$$

is known to us, where $\varrho(\xi, \eta)$ runs through all Gaussian forms of the genus R to which $\varphi(\xi, \eta)$ belongs. We propose to study the asymptoticity of the number of solutions to the equation (IX.3.1) and its asymptotic geometry on this basis.

We assume that the sequence of numbers $\{a_l\}$ is "well distributed" in arithmetic progression with differences not exceeding n^α ($\alpha > 0$ being a small constant), so that we may apply the technique of the sieve of Eratosthenes to obtain an upper bound on the number of solutions of (IX.3.2) possessing certain properties. In particular, these properties will be: the existence of a decomposition of the number $\varrho(\xi, \eta) = \pi_1 \pi_2 \ldots \pi_s$ into prime factors, the existence of numbers among these factors representing each of the Gaussian forms of the genus R, and, moreover, a sufficient number of such factors. Solutions of such a genus will be called good, and other solutions will be called poor. If $Q(m)$ is the number of solutions of (IX.3.2), then in many cases we obtain that the number of poor solutions does not exceed $\frac{Q(m)}{(\ln m)^\alpha}$, where $\alpha > 0$ is a constant. Considering good solutions and the decomposition

$$\varrho(\xi, \eta) = \pi_1 \ldots \pi_s$$

where $\pi_i = \varphi_{k_i}(\xi, \eta)$ and φ_{k_i} is a quadratic form with determinant d, we can replace the representation $\pi_i = \varphi_{k_i}(\xi, \eta)$ by $\varphi'_{k_i}(\xi', \eta')$, where $\varphi'_{k_i}(\xi', \eta')$ is a form of the genus inverse to the genus of $\varphi_{k_i}(\xi, \eta)$. We shall write such a substitution symbolically as $\pi_i = \varphi_{k_i} \to \pi_i = \varphi'_{k_i}$.

Such replacements, in numbers 2^s, can be thought of as wanderings along a trajectory and possess elementary ergodic properties; the equation $\varrho(\xi, \eta) = \varphi(\xi, \eta)$ may be considered as a transformation law for such wanderings; in view of this it is possible to deduce an asymptotic relation for the number of solutions of (IX.3.1) from such elementary ergodic considerations.

B. M. Bredihin and the author used the approach described above to study the asymptoticity of the number of solutions to the generalized Hardy-Littlewood problem (for definition of this problem see Linnik [8], Chapter IX). Here a_m runs through a sequence of prime numbers p, and thus the equation (IX.3.1) has the form

$$p + \varphi(\xi, \eta) = m\,. \tag{IX.3.3}$$

Choosing as a_l a sequence of values of another binary quadratic form $Q(x, y)$, we obtain the problem concerning the asymptotic number of representations of m by quaternary quadratic forms; here the coefficients of $Q(x, y)$ may be taken very large, so that we obtain certain results inaccessible by ordinary methods, Choosing $a_l = al^2$ we obtain the problem of representation of m by ternary forms.

The statements above become clearer if we do not investigate "wanderings on the class of quadratic forms", but instead wanderings of geometric type.

We consider the equation

(IX.3.4) $$a_l + x^2 + y^2 = m .$$

Suppose that the asymptotic number of its solutions $Q(n)$ is known. Select inside of the unit circle $x^2 + y^2 \leqq 1$ a convex region Ω containing 0 and having area $\mu(\Omega)$ and perform a homothetic stretching of Ω in the ratio $\sqrt{n}:1$, obtaining a new region $\Omega_{\sqrt{n}}$. For which conditions concerning the sequence $\{a_n\}$ is it then possible to assert that the number of solutions $Q(m, \Omega_1)$ for which the point $x + iy \in \Omega_{\sqrt{m}}$ will have the asymptotic expression

(IX.3.5) $$Q(m, \Omega_1) \sim \mu(\Omega)\, Q(n)\,?$$

In BREDIHIN-LINNIK, (IX.3.5) was deduced for the case $a_l = p$ (the Hardy-Littlewood equations). Indeed, for the equation

(IX.3.6) $$p + x^2 + y^2 = m$$

the following theorem is valid.

Theorem IX.3.1:

(IX.3.7) $$Q(m, \Omega_1) \sim \mu(\Omega)\, Q(m) \sim \mu(\Omega)\, A_0 \Pi_m \frac{m}{\ln m},$$

where

$$A_0 = \prod_p \left(1 + \frac{X_4(p)}{p(p-1)}\right),$$

$$\Pi_m = \prod_{p|m} \frac{(p-1)(p - X_4(p))}{p^2 - p + X_4(p)} .$$

We explain in this case the application of the sieve of Eratosthenes and the elementary ergodic considerations. We consider equation (IX.3.4) and form for each solution the pair (a_l, z), where $z = x + iy$. Divide the disc $|z| \leqq \sqrt{m}$ into a large but fixed number M of sectors S_k with angles $\frac{2\pi}{M}$ (the first sector having the x-axis as a boundary). We call the solution (a_l, z) good, if

(IX.3.8) $$z = x + iy = \pi_1 \pi_2 \dots \pi_s q ,$$

where π_j are Gaussian primes and among the s quantities π_j a sufficiently large number of them lie in any predetermined S_k. If we make 2^s substitutions in the good solutions of the form

$$\pi_1 \to \bar{\pi}_1, \quad \pi_2 \to \bar{\pi}_2, \quad \text{etc.},$$

then z will go over into the points $z', z'', \ldots$, and obviously $(a_l, z'), (a_l, z''), \ldots$ are new solutions having elementary ergodic properties. It is easily shown that the points $z', z'', \ldots$ possess ergodic behavior with respect to their transformations through the different sectors S_k. We cannot assert this for the poor solutions, but with the help of the sieve of Eratosthenes it is possible in many cases to show that there are relatively few poor solutions; the overwhelming number of solutions are good. From this it is not difficult to deduce the relation (IX.3.5).

For the Hardy-Littlewood equation (IX.3.6) the number of poor solutions is of the order

$$0\left(\frac{Q(m)}{(\ln m)^{\alpha_M}}\right),$$

where $\alpha_M > \dfrac{1}{3M}$ (see Bredihin-Linnik), which reduces to Theorem IX.3.1.

We turn to the case where $a_l = l^2$; setting $l = v$, we arrive at the problem of the asymptoticity of integral points on the sphere $Sp_3(m)$

(IX.3.9) $$x^2 + y^2 + v^2 = m$$

considered in Chapter IV. The number of solutions $Q(m)$ is given by the equations (IV.2.4); for the proofs in Chapter IV the lower bound of K. L. Siegel

$$Q(m) > c(\varepsilon)\, m^{\frac{1}{2}-\varepsilon}$$

was sufficient. For the method given above, however, this is not sufficient, and we need the hypothesis

(IX.3.10) $$Q(m) > \frac{m^{1/2}}{(\ln m)^{\alpha_0}},$$

where $\alpha_0 > 0$ is sufficiently small.

We consider the series $L(s, X)$ introduced in § 2. If the Riemann hypothesis is satisfied for this series, then it can be shown that

(IX.3.11) $$Q(m) > \frac{c_0 m^{1/2}}{\ln \ln m},$$

which is quite sufficient. The estimate (IX.3.10) even follows from the assumption that $L(s, X)$ has no zeros in the semicircle $|s-1| \leqq \dfrac{1}{\ln \ln m}$; $\sigma < 1$ (for sufficiently large m). Here the number α_0 in the estimate (IX.3.10) may be taken as small as desired. Let us assume that the estimate

(IX.3.10) is valid. We write the solutions (x, y, v) of equation (IX.3.9) in the form (v, z), where $z = x + iy$. We introduce the sectors S_k and determine the good solutions as described above (see BREDIHIN-LINNIK). Then the number of poor solutions is of the order

$$0\left(\frac{Q(m)}{(\ln m)^{\alpha_M - \alpha_0}}\right), \tag{IX.3.12}$$

deduced with the help of the sieve of Eratosthenes.

Since α_0 can be chosen as small as desired (important is only $\alpha_0 < \alpha_M$), then an overwhelming number of solutions will be good and the points $z = x + iy$ will be asymptotically uniformly distributed in the sectors S_k. Hence it is not difficult to deduce the asymptotic distributions of integral points (x, y, v) on the segments. It is true that the remainder term is still worse than the one obtained by the method of Chapter IV; it is of the order

$$0\left(\frac{Q(m)}{(\ln \ln m)^{1/3}}\right).$$

Analogously the condition

$$h(-D) > \frac{D^{1/2}}{(\ln D)^{\alpha_0}}, \tag{IX.3.13}$$

where $-D$ is a fundamental discriminant, reduces for sufficiently small α_0 on the basis of elementary ergodic considerations Theorem V.2.1 on the asymptotic distribution of reduced binary quadratic forms in connection with Lobachevskian geometry. Here it is sufficient to remark that the basic equation

$$ac - b^2 = D \qquad (D > 0,\ a > 0) \tag{V.1.1}$$

goes over into the form $\xi^2 - \eta^2 - b^2 = D$ by the substitutions $a = \xi - \eta$, $c = \xi + \eta$, $l = b$, and is similar to (IX.3.9) with respect to the above considerations (see BREDIHIN-LINNIK).

We note that (IX.3.11) follows from (IX.3.13) on the basis of GAUSS' theorem (Theorem III.1.1). However, (IX. 3.13) has not been proved at the present time.

Chapter X

An Arithmetic Simulation of Brownian Motion

§ 1. General Remarks. Formulation of the Problem

The present chapter stands alone; it is concerned solely with probabilistic number theory as opposed to the preceding chapters, in which we used ergodic concepts and limit theorems of probability theory for calculation of large deviations and investigation of individual asymptotic problems in the theory of diophantine equations. However, the investigation of objects having to do with the theory of quadratic fields and at the same time stochastic processes, as given in this chapter, deserves a place in this book.

This chapter is a study of the results presented in KUBILIUS-LINNIK.

To make use of the Monte Carlo method it is often necessary to simulate trajectories of Brownian motion. Here we describe one method for such a construction.

Let p be an odd prime number. We denote by (m/p) the Legendre symbol, which, as is well known, is defined for all integers m in the following manner. We set

$$\left(\frac{0}{p}\right)=0, \quad \left(\frac{k^2}{p}\right)=1 \quad \left(k=1,2,\ldots\frac{p-1}{2}\right),$$

and require that (m/p) as a function of m be periodic with period p. For all m for which the value of the symbol (m/p) is not defined by the preceding conditions, we set $(m/p)=-1$.

We consider the sum

$$S_p(m,t;h)=\frac{1}{\sqrt{h}}\sum_{n\leqq ht}\left(\frac{m+h}{p}\right),$$

where m is an integer, $h>0$, $t\geqq 0$ are real numbers, and the sum is taken over all natural $n\leqq ht$. By $N_u\{\ldots\}$ we denote the number of counting numbers $m\leqq u$ which satisfy the conditions given inside of the brackets.

DAVENPORT and ERDÖS have shown that for $p\to\infty$, $h=h(p)\to\infty$, $\dfrac{\log h}{\log p}\to 0$, the sums $S_p(m,1;h)$ $(1\leqq m\leqq p)$ are asymptotically distributed according to the normal law:

$$\frac{1}{p}N_p\{S_p(m,1;h)<x\}\to\frac{1}{\sqrt{2\pi}}\int_{-\infty}^{x}e^{-\frac{u^2}{2}}\,du\,.$$

A generalization of the methods used by DAVENPORT and ERDÖS permits the proof of a more general statement, namely, that the sums $S_p(m, t; h)$ under the same conditions and in the same sense represent in the limit Brownian motion with time $t \geqq 0$; the increments of the sums $S_p(m, t; h) - S_p(m, s; h)$ for $0 \leqq s < t$ turn out to be asymptotically normal with parameters 0, $\sqrt{t-s}$, and the increments $S_p(m, t_j; h) - S_p(m, s_j; h)$ for $0 \leqq s_j < t_j$ $(j = 1, \ldots, k)$ are asymptotically independent for any finite choice of disjoint intervals $(s_1, t_1), \ldots, (s_k, t_k)$.

The above statements can be generalized to the case of composite p. For simplicity we restrict ourselves to the case of odd square-free numbers. Let $P > 1$ be such a number. We introduce the Jacobi symbol

$$(m/P) = \prod_{p|P} (m/p)$$

where p runs through all prime divisors of P and (m/p) is the Legendre symbol.

We then set

$$S_p(m, s, t; h) = \frac{1}{\sqrt{h}} \sum_{hs < n \leqq nt} \left(\frac{m+n}{P}\right), \quad 0 \leqq s < t.$$

§ 2. Basic Theorems

Theorem X.2.1: *If P runs through any infinite increasing sequence of odd square-free numbers such that for every fixed $c \geqq 0$*

$$\prod_{p|P} (1 - c/p) \to 1 \tag{X.2.1}$$

as $P \to \infty$, $h = h(P) \to \infty$, $\dfrac{\log h}{\log P} \to 0$, then

$$\frac{1}{P} N_P(S_P(m, s, t; h) < x) \to \frac{1}{\sqrt{2\pi(t-s)}} \int_{-\infty}^{x} e^{-\frac{u^2}{2(t-s)}} du \tag{X.2.2}$$

and for any choice of disjoint intervals $(s_1, t_1), \ldots, (s_k, t_k)$, $0 \leqq s_j < t_j$ $(j = 1, \ldots, k)$,

$$\begin{aligned} &\frac{1}{P} N_P\{S_P(m, s_1, t_1; h) < x_1, \ldots, S_P(m, s_k, t_k; h) < x_k\} \to \\ &\to \prod_{j=1}^{k} \lim_{P\to\infty} \frac{1}{P} N_P\{S_P(m, s_j, t_j; h) < x_j\}. \end{aligned} \tag{X.2.3}$$

An analogous result is valid for characters of higher orders (the information we need concerning characters may be found in CHUDAKOV).

Theorem X.2.2: *Let P run through an infinite increasing sequence of odd square-free integers, occurring as basic modula of primitive characters* *of order* $g > 2$, *for which the condition* (X.2.1) *is satisfied. Let* $h = h(P)$ *satisfy the conditions of Theorem X.2.1 and*

$$T_P(m, s, t; h, \chi) = U_P(m, s, t; h, \chi) + i V_P(m, s, t; h, \chi) = \\ = \sqrt{2/h} \sum_{hs < n \leqq ht} \chi_P(m+n),$$

where $\chi_P(m)$ *is any primitive character modulo* P *of order* g. *Then the formulas* (X.2.1) *and* (X.2.2) *are valid if we replace* $S_P(m, s, t; h)$ *by* $U_P(m, s, t; h, \chi)$ *or* $V_P(m, s, t; h, \chi)$.

In particular, if $P = p$ is an odd prime with $p \equiv 1 \bmod 3$, then $U_P(m, s, t; h, \chi)$ can be determined in the following manner. We introduce the symbol $\{m/p\}$ for all rationally integral m. We set $\{0/p\} = 0$, $\{k^3/p\} = 1$ $(k = 1, 2, \ldots, p-1)$ and specify that the symbol $\{m/p\}$ should have period p as a function of m. For those symbols $\{m/p\}$ not defined by the above regulations, we set $\{m/p\} = -\frac{1}{2}$. Then

$$U_P(m, s, t; h, \chi) = \sqrt{\frac{2}{h}} \sum_{hs < n \leqq ht} \left\{\frac{m+n}{p}\right\}.$$

For the calculation of Legendre and Jacobi symbols there exist many algorithms suitable for programming. For example, we can make use of the following properties of the Jacobi symbol (P, P_1, P_2 are positive odd numbers, m, m_1, m_2 integers):

1) $\left(\frac{m^2}{P}\right) = 1$ if $(m, P) = 1$,

2) $\left(\frac{m}{P}\right) = 0$ if $(m, P) > 1$,

3) $\left(\frac{m_1}{P}\right) = \left(\frac{m_2}{P}\right)$ if $m_1 \equiv m_2 \bmod P$,

4) $\left(\frac{m_1 m_2}{P}\right) = \left(\frac{m_1}{P}\right)\left(\frac{m_2}{P}\right)$,

5) $\left(\frac{m}{P_1 P_2}\right) = \left(\frac{m}{P_1}\right)\left(\frac{m}{P_2}\right)$,

6) $\left(\frac{-1}{P}\right) = (-1)^{\frac{P-1}{2}}$

7) $\left(\frac{2}{P}\right) = (-1)^{\frac{P^2-1}{8}}$

8) $\left(\frac{Q}{P}\right) = (-1)^{\frac{P-1}{2}\frac{Q-1}{2}}\left(\frac{P}{Q}\right)$,

where Q is a positive odd integer relatively prime to P. In this way, $(-1/P) = 1$ if $P \equiv 1 \pmod 4$ and $(-1/P) = -1$ if $P \equiv -1 \pmod 4$; $(-2/P) = 1$ if $P \equiv \pm 1 \pmod 8$ and $(-2/P) = -1$ if $P \equiv \pm 3 \pmod 8$; $(Q/P) = (P/Q)$ if

$P\equiv 1\pmod 4$ or $Q\equiv 1\pmod 4$, and $(Q/P)=-(P/Q)$ otherwise. The application of the following algorithm is highly suitable, too.

Let P and Q be odd relatively prime positive integers. We form the sequence

$$Q=P_q+P_1,\ P=|P_1|\,q_1+P_2,\ |P_1|=|P_2|\,q_2+P_3,\ |P_2|=|P_3|\,q_3+P_4,\ \dots,\ |P_{k-1}|=|P_k|\,q_k+P_{k+1},$$

where $P_0=P, P_1, P_2, \dots, P_{k+1}=1$ are positive or negative integers decreasing in absolute value; $q, q_1, q_2, \dots, q_k$ are even numbers greater than or equal to 2. If the number of ν ($\nu=0, 1, \dots, k$) for which $|P_\nu|$ and $|P_{\nu+1}$ are simultaneously congruent $-1 \bmod 4$ is α, then

$$(Q/P)=(-1)^\alpha.$$

For the calculation of the characters $\chi_P(m)$ there also exist certain algorithms.

We pass to the proofs of the theorems. We only prove Theorem X.2.2. Theorem X.2.1 is proved analogously, only more simply.

We consider the moments of the distribution function of the sum $T_p(m, s, t; h, \chi)$:

$$\mu_{qr}(P)=\frac{1}{P}\sum_{m=1}^{P} T_P^q(m, s, t; h, \chi)\,\bar{T}_P^r(m, s, t; h, \chi),$$

where q, r are non negative integers and the bar denotes complex conjugation. Since $\mu_{00}(P)=1$, we may restrict ourselves to the case $q+r\geqq 1$. In view of the complete multiplicative property of the character $\chi_P(m)$ and respectively $\bar{\chi}_P(m)=\chi_P(m^{q-1})$, we have

$$\mu_{qr}(P)=\left(\frac{1}{P}\right)\frac{2^{\frac{q+r}{2}}}{h}\sum_{n_1}\dots\sum_{n_{q+r}}\sum_{m=1}^{P}\chi_P((m+n_1)\dots \\ \dots(m+n_q)(m+n_{q+1})^{q-1}\dots(m+n_{q+r})^{q-1}), \tag{X.2.4}$$

where $n_1, \dots, n_{q+r}$ run independently of each other over all counting numbers in the interval (hs, ht).

We denote the polynomial in the parenthesis after χ_P in (X.2.4) by $f(m)$ and consider the sum

$$\sum_{m=1}^{P}\chi_P(f(m)).$$

Let $P=p_1p_2\dots p_l$ be the factorization of P into a product of primes with the cofactors

$$P_j=P/p_j \qquad (j=1, \dots, l).$$

Then

$$\sum_{m=1}^{P}\chi_P(f(m))=\sum_{m_1=1}^{P_1}\dots\sum_{m_l=1}^{P_l}\chi_P(f(m_1P_1+\dots+m_lP_l)).$$

The character $\chi_P(m)$ can be represented in the form of a product

$$\chi_{p_1}(m)\dots\chi_{p_l}(m)$$

of primitive characters modulo $p_1, \ldots, p_l$. We have

$$\sum_{m=1}^{P} \chi_P(f(m)) = \prod_{j=1}^{l} \sum_{m_j=1}^{p_j} \chi_{p_j}(f(m_1 P_1 + \cdots + m_l P_l)) =$$

$$= \prod_{j=1}^{l} \sum_{m=1}^{p_j} \chi_{p_j}(f(m P_j)) = \prod_{j=1}^{l} \sum_{m=1}^{p_j} \chi_{p_j}(f(m)).$$

We estimate the preceding sum for $f(m)$ being the g-th power of a polynomial. Since the degree of the polynomial $f(m)$ is equal to $r + \frac{q-r}{g}$ (in this case $q-r$ is divisible by g), the congruence $f(m) \equiv 0 \pmod{p_j}$ has no more than $r + \frac{q-r}{g}$ solutions. Consequently,

$$\sum_{m=1}^{p_j} \chi_{p_j}(f(m)) = p_j - \theta\left(r + \frac{q-r}{g}\right)$$

where $0 \leqq \theta \leqq 1$. By virtue of condition (X.2.1) we have

$$\text{(X.2.5)} \qquad \sum_{m=1}^{P} \chi_P(f(m)) = \prod_{j=1}^{r} \left(p_j - \theta\left(r + \frac{q-r}{g}\right)\right) = P(1 + o(1)).$$

If $f(m)$ is not the g-th power of a polynomial, then we obtain from results due to A. Weil ([1], [2])

$$\left|\sum_{m=1}^{p_j} \chi_{p_j}(f(m))\right| < C\sqrt{p_j},$$

where C depends only on q, r, and g. In this manner,

$$\left|\sum_{m=1}^{P} \chi_P(f(m))\right| < C^l \sqrt{P}.$$

Let p_0 be the smallest prime divisor of P. Then $p_0^l \leqq P$, whence $l \leqq \frac{\log P}{\log p_0}$. By virtue of condition (X.2.1) p_0 increases unboundedly together with P. Therefore for sufficiently large P

$$\log C < \tfrac{1}{4} \log p_0 \quad \text{and} \quad C^l < P^{\frac{1}{4}},$$

consequently

$$\text{(X.2.6)} \qquad \sum_{m=1} \chi_P(f(m)) < P^{\frac{3}{4}}.$$

We return now to the formula (X.2.4). Let Q_{qr} and Q'_{qr} denote respectively the number of systems of numbers $n_1, \ldots, n_{q+r}$ for which $f(m)$ is or is not the g-th power of a polynomial, where the numbers $n_1, \ldots, n_{q+r}$ run through all counting numbers in the interval (hs, ht) independently of each other. Then by virtue of the estimates (X.2.5) and (X.2.6) we have

$$\text{(X.2.7)} \qquad \mu_{qr}(P) = (2/h)^{\frac{q+r}{2}} Q_{qr}(1 + o(1)) + 0(P^{-\frac{1}{4}} h^{-\frac{q+r}{2}} Q'_{qr}).$$

Trivially we have

$$Q'_{qr} < (h(t-s)+1)^{q+r} = 0(h^{q+r}).$$

We restrict ourselves to the case $r \leqq q$. It is not difficult to compute that

$$Q_{qr} < C_1 (h(t-s)+1)^{r+\frac{q-r}{g}} = 0(h^{r+\frac{q-r}{g}}),$$

where C_1 is dependent only on q, r, and g. It follows from these estimates that

$$\mu_{qr}(P) = 0(h^{-(q-r)(\frac{1}{2}-\frac{1}{g})} + P^{-\frac{1}{4}} h^{\frac{q+r}{2}}),$$

whence, according to the condition $\dfrac{\log h}{\log P} \to 0$, we obtain

$$\mu_{qr}(P) \to 0 \quad \text{for } r < q \text{ and } P \to \infty .$$

One proves that $\mu_{qr}(P) \to 0$ for $q < r$ completely analogously.

For the estimate of $\mu_{qr}(P)$ where $q = r$ we compute Q_{qq} exactly. First of all, the number of systems $n_1, \ldots, n_{2q}$ for which $f(m)$ is a g-th power of a polynomial and among which there exist more than q different n_j is less than

$$(h(t-s)+1) \cdot q \cdot (h(t-s)+1)\,(q-1) \ldots (h(t-s)+1) \cdot 1 .$$

In this manner,

$$Q_{qq} < q!(h(t-s)+1) = q!\, h^q (t-s)^q\, (1 + 0(1/h)).$$

Secondly, the number of systems $n_1, \ldots, n_{2q}$ for which $f(m)$ is a g-th power of a polynomial and among which there exist exactly q different numbers, where all $n_1, \ldots, n_q$ are different from each other, is not less than

$$[h(t-s)] \cdot q \cdot [h(t-s)-1] \cdot (q-1) \ldots [h(t-s)-q+1] \cdot 1 .$$

Hence it follows that

$$Q_{qq} \geqq q!\, h^q (t-s)^q\, (1 + 0(1/h)).$$

Finally we have

(X.2.8) $$Q_{qq} = q!\, h^q (t-s)^q (1 + 0(1/h)).$$

From (X.2.7) and (X.2.8) we conclude that

$$\mu_{qq}(P) = 2^q q!\, (t-s)^q\, (1 + 0(1)) + 0(P^{-\frac{1}{4}} h^q) \to 2^q q!\, (t-s)^q$$

as $P \to \infty$.

It is not difficult to see that the limits of the moments $\mu_{qr}(P)$ are equal to the corresponding moments of the complex random variable $\zeta = \xi + i\eta$ with distribution function

$$P_r\{\xi < x, \eta < y\} = \frac{1}{2\pi(t-s)} \int_{-\infty}^{x} \int_{-\infty}^{y} e^{-\frac{u^2+v^2}{2(t-s)}}\, du\, dv .$$

Indeed,

$$M(\zeta^q \bar{\zeta}^r) = \frac{1}{2\pi(t-s)} \int_{-\infty}^{\infty} \int_{-\infty}^{\infty} (u+iv)^q (u-iv)^r e^{-\frac{u^2+v^2}{2(t-s)}}\, du\, dv .$$

Introducing the change of coordinates

$$u=\varrho\sqrt{t-s}\cos\varphi\,,\quad v=\varrho\sqrt{t-s}\sin\varphi$$

we get

$$M(\zeta^q\bar{\zeta}^r)=\frac{1}{2\pi}(t-s)^{\frac{q-r}{2}}\int\limits_0^\infty \varrho^{q+r+1}e^{-\frac{\varrho^2}{2}}\,d\varrho\int\limits_0^{2\pi}e^{-i(q-r)\varphi}\,d\varphi\,.$$

The interior integral is equal to zero for $q\neq r$ and equal to 2π for $q=r$. In addition,

$$\int\limits_0^\infty \varrho^{2q+1}e^{-\frac{\varrho^2}{2}}\,d\varrho=2^q q!\,.$$

Thus

$$M(\zeta^q\bar{\zeta}^r)=\begin{cases}0 & \text{for } q\neq r\\ 2^q(t-s)^q q! & \text{for } q=r\,.\end{cases}$$

From all of the preceding it follows that the sums $U_P(m, s, t; h, \chi)$ and $V_P(m, s, t; h, \chi)$ are asymptotically normally distributed with mean value 0 and standard deviation $t-s$. Here we have also proved that U and V are asymptotically indepedent.

For the completion of the proof to the theorem it remains to be shown that for any choice of non-negative integers $q_1, r_1, \dots, q_k, r_k$ the moments

$$\mu_{q_1r_1,\dots,q_kr_k}(P)=\frac{1}{P}\sum_{m=1}^{P}T_P^{q_1}(m, s_1, t_1; h, \chi)\times$$
$$\times\,\bar{T}_P^{r_1}(m, s_1, t_1; h, \chi)\times\cdots$$
$$\cdots\times T_P^{q_k}(m, s_k, t_k; h, \chi)\,\bar{T}_P^{r_k}(m, s_k, t_k; h, \chi)$$

for disjoint intervals $(s_1, t_1), \dots, (s_k, t_k)$ approach the corresponding moments of the complex random vector

$$(\xi_1+i\eta_1, \dots, \xi_k+i\eta_k)$$

with the distribution function

$$P_r\{\xi_1<x_1, \eta_1<y_1, \dots, \xi_k<x_k, \eta_k<y_k\}=$$
$$=\prod_{j=1}^{k}\frac{1}{2\pi(t_j-s_j)}\int\limits_{-\infty}^{x_j}\int\limits_{-\infty}^{y_j}e^{-\frac{u^2+v^2}{2(t_j-s_j)}}\,du\,dv\,.$$

The proof is completely analogous to the exposition above for the special case $k=1$.

Analogous results may also be obtained for the functions

$$\chi(F(m)),\ \exp\left(\frac{2\pi i}{P}F(m)\right),\ \chi(F_1(m))\exp\left(\frac{2\pi i}{P}F(m)\right).$$

where $F(m)$ and $F_1(m)$ are integral polynomials and $\chi(m)$ a Dirichlet character.

Chapter XI

Supplementary Remarks. Problems

As usual, the unsolved problems belonging to the given region of investigation are much more numerous than the solved ones, and they are much more difficult. Here we state some unsolved problems of interest and make some supplementary remarks concerning the other chapters.

Chapter II, § 1. Certain unsolved problems are present in the arithmetic of hermitions; we refer to the important works in recent years of M. EICHLER, and, in the direction of this book, the works of A. V. MALYSHEV. The existence of the Euclidean algorithm is not entirely resolved in the arithmetic of indeterminate hermitions; many unsolved problems remain in the arithmetic of hermitions over arbitrary algebraic number fields (which is also of interest for Chapter IX, § 1).

Chapter III. The constructions and estimates connected with rotations of a large sphere are cumbersome and the estimates are gross. A simplification of the estimation methods is desirable and more exact estimates can probably be obtained. For possibilities see LINNIK-VINOGRADOV.

Chapters IV, V, VI. The ergodic and mixing theorems presented in these chapters are undoubtedly of such a character as to justify these names. However, their proof purely by means of ergodic theory does not succeed, if only because the set of integral points is finite and has Lebesgue measure zero. A situation would be highly desirable in which these theorems not only are formulated using the concepts of ergodic theory, but also proved by these means, so that the entire investigation would have a certain place in ergodic theory.

Chapter IV. Under what conditions may Theorem IV.1.1 be formulated without the auxiliary number q? What can be done in this direction with the generalized Riemann hypothesis and its weaker forms, aside from that already known (see A. V. MALYSHEV, pp. 179—190)?

Is it possible to assume that the solid angle $\omega(\Gamma)$ converges to 0 together with $1/m$ and how fast is this convergence? (From the method presented it follows that it is possible to take $\omega(\Gamma) \asymp \frac{1}{(\ln m)^\alpha}$ for some $\alpha > 0$.) On the sphere with radius $\sqrt{m}$ we know, roughly speaking, that the number of integral points is of the order $\sqrt{m}$, and it may be expected that the solid angle $\omega(\Gamma)$ is of the order $m^{-\frac{1}{2}+\varepsilon_0}$ for any $\varepsilon_0>0$, so that for any convex region on the sphere the area would be larger than $c(\varepsilon)m^{\frac{1}{2}+\varepsilon}$; then it could be expected that primitive integral points exist and are asymptotically uniformly distributed. But how do we prove these statements? In particular, is the equation

$$x^2+y^2+z^2=m$$

solvable for $x=0(m^{\varepsilon})$? We not only do not know the answer to this, but we also do not know whether the diophantine unequality

$$|m-y^2-z^2|<m^{\varepsilon} \tag{XI.1}$$

is solvable for arbitrarily large m. How much can we improve the remainder term κ_0 in Theorem IV.1.1 (relation (IV.1.2))? The given method results only in

$$\kappa_0=0\left(\frac{1}{(\ln m)^{\alpha}}\right)$$

for a certain $\alpha>0$.

Chapter V, § 2. Concerning Theorem V.2.1 and V.2.2 problems of the same type as in the remark to Chapter IV, § 1, arise. Of special importance for the theory of Dirichlet characters is the problem whether in Theorem V.2.3 we may put $\varepsilon=\dfrac{1}{D^{\frac{1}{2}-\eta}}$, where η is as small as desired. This is possible if the Riemann hypothesis is true for the series $L(s,X)$.

Chapter V, § 19. It is of interest to deduce ergodic theorems for modular invariants without the use of quadratic forms and matrices by investigating the corresponding "fractional parts" (see Chapter V, § 19) or by some other method.

Chapter VI, § 2. The theorem on cycles (Theorem VI.2.2) can be formulated in terms of reduced binary indefinite forms without the help of matrices and rotations. Is it possible to prove this theorem using only the methods of the theory of such forms?

Chapter VII, § 5. What other possibilities are there for the formulation of ergodic theorems and mixing theorems? How can these be classified? How can we construct a theory for hyperboloids of one sheet of a more general nature?

Chapter VII, § 3. How can we construct a complete theory of versor rotations along the lines of Chapters IV, V, VI, and how does one prove the necessary estimates? What possibilities present themselves for the formulation of ergodic theorems and mixing theorems? For some possibilities see LINNIK-VINOGRADOV.

Do there exist connections with automorphic functions as in Chapter V, § 19? How can cubic forms be investigated using the given methods?

Chapter IX, § 1. It would be interesting to investigate a system (IX.1.3) with two quadratic forms, its asymptotic geometric solutions, and ergodic and mixing theorems. Here the wandering region is compact. How can this system be generalized to more quadratic forms?

Chapter IX, § 2. For which conditions on $L(s, X)$ in the sense of a weakened Riemann hypothesis will Theorem IV.1.1 be valid without the auxiliary number q?

Chapter IX, § 3. Applying the necessary weak Riemann hypothesis, one can then study asymptotic representations of numbers by positive ternary quadratic forms and their basic ergodic properties as in § 3. On this basis one could give new proofs of the asymptotic distribution of integral points on the sphere.

Chapter X. Does the normalized sum of values of the Möbius function $\mu(n)$ represent in the limit the Brownian motion stochastic process (Wiener process)? More precisely, consider a sequence of integers $P_k \to \infty$; substituting in Theorem X.2.1 P_k for P and $\mu(m+n)$ for $\left(\frac{m+n}{P}\right)$, is it possible to obtain an analogy of the limit expression (X.2.3)?

In the above we were interested in a positive result; a negative result would also be interesting, since in combination with Theorem X.2.1 it would give a solution to the problem of the effective calculation of ideal classes in an imaginary quadratic field.

Bibliography

BACHMANN, P.: Die Arithmetik der quadratischen Formen I. Leipzig 1898.

BERNSTEIN, S. N.: Probability theory. GITTL 1946.

BIRKHOFF, G. D.: Proof of the ergodic theorem. Proc. Nat. Acad. Sci. U.S. **17**, (1957).

BREDIHIN, B. M., and YU. V. LINNIK: Binary additive problems and ergodic properties of their solutions. Dokl. Akad. Nauk U.S.S.R. **166**, (1966).

BURGESS, D. A.: The distribution of quadratic residues and nonresidues. Math. **4**, 106—112 (1957).

CHATELET, A.: Ann. École Norm. Super. **3**, 105—202 (1911).

CHEBOTAREV, N. G.: [1] Elements of the Gaussian theory II. GITTL 1936.

— [2] Theory of Lie Groups. GITTL 1940.

CHINTCHIN, A. JA.: Continued fractions. ONTI 1935. (Kettenbrüche. Leipzig: B. G. Teubner 1956.)

CHUDAKOV, N. G.: An introduction to the theory of Dirichlet L-series. GITTL 1947.

DAVENPORT, H., and P. ERDÖS: The distribution of quadratic and higher residues. Publ. Math. **2**, 252—265 (1952).

EICHLER, M.: Quadratische Formen und orthogonale Gruppen. Berlin 1952.

GELFOND, A. O., and YU. V. LINNIK: Elementary methods of analytic number theory. FIZMATGIZ 1962.

GNEDENKO, B. V.: A course in probability theory. FIZMATGIZ 1961.

HALMOS, P.: [1] Measure theory. New York 1950.

— [2] Lectures on ergodic theory. Math. Soc. Japan 1956.

JACOBSON, N.: The theory of rings. AMS Math. Surv. N.Y. 1943.

KAC, M.: Statistical independence in probability, analysis, and number theory. New York 1952.

KLOOSTERMAN, H. D.: On the representation of numbers in the form $ax^2 + by^2 + cz^2 + dt^2$. Acta Math. **49**, (1926).

KUBILIUS, I. P., and YU. V. LINNIK: Arithmetic models of Brownian motion. IZV. VUZOV. MAT. **6**, 88—95 (1959).

KUZMIN, R. O.: On a problem of Gauss. Dokl. Akad. Nauk Series A, 375—380 (1928).

LINNIK, YU. V.: [1] On representations of large numbers by positive ternary quadratic forms. Izv. Akad. Nauk **4**, 363—402 (1940).

— [2] Quaternions and Cayley numbers. Some applications of quaternion arithmetic. Usp. **4**, 5 (33), 49—98 (1949).

— [3] The asymptotic distribution of reduced binary quadratic forms in connection with Lobachevskian geometry. Vestn. Leningr. Univ. Ser. Mat. Nr. 2, 3—23 (1955), Nr. 5, 3—32 (1955), Nr. 8, 15—27 (1955).

— [4] Markov chains in the analytic arithmetic of quaternions and matrices. Vestn. Leningr. Univ. Ser. Mat. Nr. 13, 63—68 (1956).

— [5] Other analogies of ergodic theorems for imaginary quadratic fields. Dokl. Akad. Nauk U.S.S.R. **109**, 694—696 (1956).

— [6] Asymptotic-geometrical and ergodic properties of sets of integral points on the sphere. Math. Sbornik **43** (85), 2, 257 (1957).

— [7] Some applications of non-euclidean geometry to the theory of Dirichlet characters; analogies to ergodic theorems. Proc. 3rd All-Union Math. Cong. Vol. III (1958).

— [8] A dispersion method for binary additive problems. Leningrad 1961.

LINNIK, YU. V.: [9] Additive problems and eigenvalues of the modular operators. Proc. Int. Cong. Math. Stockholm, 280—284 (1963).

—, and A. V. MALYSHEV: Applications of quaternion arithmetic to the theory of ternary quadratic forms and to the analysis of numbers on cubes. Usp. VIII, 5 (57), 3—71 (1953) [correction in Uspehi I (1955)].

—, and B. F. SKUBENKO: Asymptotic distribution of integral matrices of the third order. Vestn. Leningr. Univ. Ser. Mat. Nr. 13, 25—36 (1964).

MALYSHEV, A. V.: On representations of integers by positive quadratic forms. (V. A. STEKLOVA) Trudy Math. Inst. Akad. Nauk **45**, (1962).

POMMERENKE, G.: Über die Gleichverteilung von Gitterpunkten auf m-dimensionalen Ellipsoiden. Acta Arith. **5**, 227—257 (1959).

RYLL-NARDZEWSKY, C.: On the ergodic theorem II. Studia Math. Vol. 12, 74—79 (1951).

SCHUR, J.: Sitzber. Preuss. Akad. Wiss. **13—14**, 145—168 (1922).

SIEGEL, C. L.: Über die Klassenzahl quadratischer Zahlkörper. Acta Arith. **1**, 83—86 (1935).

SKUBENKO, B. F.: [1] The asymptotic distribution of integral points on the hyperboloid of one sheet and ergodic theorems. Izv. Akad. Nauk **26**, 721—752 (1962).

— [2] On the asymptoticity of integral matrices of the n-th order. Dokl. Akad. Nauk U.S.S.R. **159**, 290—291 (1963).

SYLVESTER, J. J.: Sur l'équation en matrices $px=xq$. C. R. Acad. Sci Paris **19**, 67—71, 115—116 (1884).

USPENSKI, YA. V.: An introduction to the non-euclidean geometry of Lobachevsky-Bolian. PG 1922.

VENKOV, B. A.: [1] On the arithmetic of quaternions. Izv. Akad. Nauk 205—220 (1922), 221—246 (1922), 489—504, 532—562, 607—622 (1929).

— [2] Elementary number theory. ONTI 1937.

— [3] On indeterminate quadratic forms with integral coefficients. Trudy Akad. Nauk **38**, 30—41 (1951).

VINOGRADOV, I. M.: [1] Selected works. Doklady Akad. Nauk UdSSR (1952).

— [2] Basic number theory. GITTL 1952.

—, and YU. V. LINNIK: Hyperelliptic curves and least prime quadratic residues. Dokl. Akad. Nauk **168**, (1966).

WEBER, H.: Lehrbuch der Algebra III. Elliptische Funktionen und algebraische Zahlen. Braunschweig 1908.

WEIL, A.: [1] Sur les courbes algébriques et les variétés qui s'en déduisent. Actual Math. Sci. No. 1041 (1945).

— [2] On some exponential sums. Proc. Nat. Acad. Sci. U.S. **34**, 204—207 (1948).

Author Index

Page numbers in *italics* refer to the bibliography

Subject Index

Ergebnisse der Mathematik und ihrer Grenzgebiete

Lieferbare Bände:

1. Bachmann: Transfinite Zahlen. DM 38,—; US $ 9.50
4. Samuel: Méthodes d'Algèbre Abstraite en Géométrie Algébrique. DM 26,—; US $ 6.50
5. Dieudonné: La géométrie des groupes classiques. DM 38,—; US $ 9.50
6. Roth: Algebraic Threefolds with Special Regard to Problems of Rationality. DM 19,80; US $ 4.95
7. Ostmann: Additive Zahlentheorie. 1. Teil: Allgemeine Untersuchungen. Unveränderter Nachdruck von 1956. DM 38,—; US $ 9.50
8. Wittich: Neuere Untersuchungen über eindeutige analytische Funktionen. DM 28,—; US $ 7.00
11. Ostmann: Additive Zahlentheorie. 2. Teil: Spezielle Zahlenmengen. DM 22,—; US $ 5.50
13. Segre: Some Properties of Differentiable Varieties and Transformations. With Special Reference to the Analytic and Algebraic Cases. DM 36,—; US $ 9.00
14. Coxeter/Moser: Generators and Relations for Discrete Groups. DM 32,—; US $ 8.00
16. Cesari: Asymptotic Behavior and Stability Problems in Ordinary Differential Equations. DM 36,—; US $ 9.00
17. Severi: Il teorema di Riemann-Roch per curve. Superficie e varietà questioni collegate. DM 23,60; US $ 5.90
18. Jenkins: Univalent Functions and Conformal Mapping. DM 34,—; US $ 8.50
19. Boas/Buck: Polynomial Expansions of Analytic Functions. DM 16,—; US $ 4.00
20. Bruck: A Survey of Binary Systems. DM 36,—; US $ 9.00
21. Day: Normed Linear Spaces. DM 17,80; US $ 4.45
22. Hahn: Theorie und Anwendung der direkten Methode von Ljapunov. DM 28,—; US $ 7.00
24. Kappos: Strukturtheorie der Wahrscheinlichkeitsfelder und -räume. DM 21,80; US $ 5.45
25. Sikorski: Boolean Algebras. DM 38,—; US $ 9.50
26. Künzi: Quasikonforme Abbildungen. DM 39,—; US $ 9.75
27. Schatten: Norm Ideals of Completely Continuous Operators. DM 23,60; US $ 5.90
28. Noshiro: Cluster Sets. DM 36,—; US $ 9.00
29. Jacobs: Neuere Methoden und Ergebnisse der Ergodentheorie. DM 49,80; US $ 12.45
30. Beckenbach/Bellman: Inequalities. DM 30,—; US $ 7.50
31. Wolfowitz: Coding Theorems of Information Theory. DM 27,—; US $ 6.75
32. Constantinescu/Cornea: Ideale Ränder Riemannscher Flächen. DM 68,—; US $ 17.00
33. Conner/Floyd: Differentiable Periodic Maps. DM 26,—; US $ 6.50
34. Mumford: Geometric Invariant Theory. DM 22,—; US $ 5.50
35. Gabriel/Zisman: Calculus of Fractions and Homotopy Theory. DM 38,—; US $ 9.50

36. Putnam: Commutation Properties of Hilbert Space Operators and Related Topics. DM 28,—; US $ 7.00
37. Neumann: Varieties of Groups. DM 46,—; US $ 11.50
38. Boas: Integrability Theorems for Trigonometric Transforms. DM 18,—; US $ 4.50
39. Sz.-Nagy: Spektraldarstellung linearer Transformationen des Hilbertschen Raumes. DM 18,—; US $ 4.50
40. Seligman: Modular Lie Algebras. DM 39,—; US $ 9.75
41. Deuring: Algebren. DM 24,—; US $ 6.00
42. Schütte: Vollständige Systeme modaler und intuitionistischer Logik. DM 24,—; US $ 6.00
43. Smullyan: First Order Logic. DM 36,—; US $ 9.00
44. Dembowski: Finite Geometries. In Vorbereitung
45. Linnik: Ergodic Properties of Algebraic Fields. DM 44,—; US $ 11.00
46. Krull: Idealtheorie. DM 28,—; US $ 7.00
47. Nachbin: Topology on Spaces of Holomorphic Mappings. In Vorbereitung